Unfavorable environmental conditions induce the production of resting spores in certain organisms. Many algae have successfully developed specialized resistant characteristics, such as thickened cell walls accompanied by decreased metabolic rates, that enable them to resist periods of extreme change in their environment. These strategies give the algae considerable evolutionary advantages over organisms that are unable to withstand changes in temperature, light, or ionic conditions.

Though the resting spore is considered to be an advantageous and primitive trait, the benefits are offset by the great amount of energy needed to produce and maintain the cell in near-dormancy over long periods of time and by the potentially "lost" number of cell divisions that could have occurred during the resting phase. The interesting contrast of advantages and disadvantages has stimulated biologists to investigate the morphology and the underlying processes of the physiology of vegetative cells and thick-walled resting spores.

The chapters in this book provide an excellent background to the ecological conditions and population dynamics of both marine and freshwater algae from diverse habitats and will be particularly useful to biologists and paleoecologists.

Survival strategies of the algae

Survival strategies of the algae

Edited by

GRETA A. FRYXELL
Department of Oceanography
Texas A&M University

CAMBRIDGE UNIVERSITY PRESS
Cambridge
London New York New Rochelle
Melbourne Sydney

CAMBRIDGE UNIVERSITY PRESS
Cambridge, New York, Melbourne, Madrid, Cape Town, Singapore,
São Paulo, Delhi, Dubai, Tokyo, Mexico City

Cambridge University Press
The Edinburgh Building, Cambridge CB2 8RU, UK

Published in the United States of America by Cambridge University Press, New York

www.cambridge.org
Information on this title: www.cambridge.org/9780521180085

First published 1983
First paperback edition 2010

A catalogue record for this publication is available from the British Library

Library of Congress Cataloguing in Publication data
Main entry under title:
Survival strategies of the algae.
Papers from a symposium held in Vancouver,
B.C., July 15, 1980, which was sponsored by
the Phycological Society of America and the
Systematic and Botanical Sections of the
Botanical Society of America.
Includes bibliographies.
1. Algae – Cytology – Congresses. 2. Algae –
Spores – Congresses. I. Fryxell, Greta A.
II. Phycological Society of America.
III. Botanical Society of America. Systematic
Section. IV. Botanical Society of America.
Botanical Section.
QK565.S87 1982 589.3′16 82–12865

ISBN 978-0-521-25067-2 Hardback
ISBN 978-0-521-18008-5 Paperback

Contents

Preface

The symposium "Survival Strategies of the Algae" was held in Vancouver, British Columbia, July 15, 1980, and was sponsored by the Phycological Society of America and the Systematic and Phycological Sections of the Botanical Society of America. It provided an opportunity to gather together biologists working on Cyanobacteria (Cyanophytes), Chlorophytes, and Chrysophytes (both Chrysophyceae and Bacillariophyceae) to summarize recent work relating to the production, function, and fate of specialized cells resistant to unfavorable conditions. Often these cells have thickened walls and may have morphologies that are either similar to or markedly different from their vegetative cells. The resting stages may or may not be completely dormant, but they do differ physiologically from the vegetative stage. In some groups, they can be produced by a time- and energy-consuming sequence of determinate divisions, with actual loss of nuclear material. Residual bodies from the cytoplasm often containing cell organelles are sometimes discarded with the old vegetative theca. Schmid (1979) recently reported that the diatom *Navicula cuspidata* (Kützing) Kützing forms a complete endocell nested inside another complete plus a partial theca, probably indicating four mitotic divisions in the resting spore production. In general, germination of resting spores also appears to consist of a series of determinate divisions in a given time sequence. The fact that ungerminated cells of different groups are found in sediment indicates that the individual cell pays a high cost, not only in the energy required to make the additional nuclear divisions and produce the physiological changes from the vegetative state, but also in the "lost" divisions during a resting period, and the possibility of the cell once again being in conditions favorable for germination is finite. Since the production of resting spores or akinetes or statospores is costly to the cell, if not the population, the concept of evolutionary advantage ("strategy") is interesting to explore.

Unfavorable conditions include a plethora of physical and chemical conditions that may be seasonal: fluctuations of light, temperature, and ionic conditions; ingestion by grazers and subjection to grinding and abrasion plus digestive secretions; presence of poisons in the environment; and poor nutrient conditions, such as the period following a phytoplankton bloom. It follows that cells resistant to these changes because of altered metabolic rates, heavy cell walls, and reduced exchange with the surrounding medium, or some combination of these characters, will have an adaptive advantage.

As Coleman (Chapter 1) points out, these problems are shared by marine, freshwater, and terrestrial organisms, but the overriding problem for nonmarine organisms in their survival and dissemination is water loss. In the green algae, specialized thick-walled cells are much more common in freshwater and soil algae than in the sea. Sandgren (Chapter 2) adds that resting stages serve as refuge populations that are able to recolonize a habitat when environmental conditions will again support the growth of vegetative cells. He points out that there may be either sexual or asexual processes involved, with the morphology of the thickened cell wall being typical for a given class and similar to, or different from, the vegetative cell. Recent breakthroughs in our understandings came about because electron microscopy has allowed careful morphological and ultrastructural study of both cultures and field samples.

Although desiccation is a much greater threat to freshwater algae than to most marine algae, Hargraves and French (Chapter 3) indicate that more cases of resting spores in diatoms are reported from the marine habitat than from freshwater and terrestrial habitats. Furthermore, Dale (Chapter 4), working with fossilized as well as living dinoflagellates, deals almost entirely with marine forms, although freshwater resting spores are known (Chapman et al., 1982). The heavy-walled nature of most of the resting spores or statospores of many groups, even those of planktonic species, have an increased sinking rate in both freshwater and marine habitats. However, sinking in coastal waters or the open ocean puts the cells in a different current regime, and although vast numbers of spores are not produced by each cell, this may well serve to be a dissemination mechanism (Garrison, 1980, 1981). In other habitats, viable resting spores spend part of the annual cycle in or on the sediments.

A fine summary of recent work on akinetes of Cyanobacteria was presented at the symposium in Vancouver by Dr. Norma J. Lang, University of California, Davis, but she thought that recent published accounts covered the material very well. Readers are referred to Stanier and Cohen-Dazire (1977), Nichols and Carr (1977), Carr (1979), Wolk (1979), and Carr and Whitton (1982) for summaries of recent work and references.

Resting spore production is considered a primitive trait evolutionarily. It would appear from Dale (Chapter 4) and Hargraves and French (Chapter 3), as well as from our own work (Fryxell et al., 1981), that the evolutionary pressures on vegetative cell and resting spore morphology bring about changes at different rates over geological time. In the case of the dinoflagellates and diatoms thus far reported, the resting spore morphology has diverged to a greater extent than that of the vegetative cells. Thus, early speciation appears to be indicated by divergence of resting spore morphology. However, it may well be true of some lines of descent and not of others that vegetative forms are more conservative.

There are many problems in dealing with resting spores. Coleman (Chapter 1) points out the likelihood of a "population" of resting spores in any one habitat being of markedly different ages. Thus, a gene pool must be considered to extend both temporally and spatially. Another problem deals with correctly associating the distinctive morphology of the resting spore with that of its vegetative cell. Such matching is no problem with field samples when a "bloom" of phytoplankton has occurred, and the resting spores may be found within the vegetative cells in at least some cases. However, on many occasions only suspected resting spores are found – for instance, by geologists studying the fossil record – and biologists have the opportunity (some might call it an obligation) to understand the ecological conditions that produced these cells and the corresponding physiological changes they underwent. This opportunity was the motivation behind this symposium volume, and the authors have excelled in exploring the biology and ecology of resting spores in nature.

Reference was made to field samples of a "bloom" of phytoplankton. A clonal culture is a controlled unialgal bloom, and, like a bloom, it has millions of cells available for study, with the advantage that all are descended from one cell. The large number is needed for understanding life histories, including resting spore formation and germination. Thus, there is no uncertainty involved in relating vegetative and resting stages of the same species. In addition, "foolproof" samples that require little searching are at a premium when working on an expensive instrument, such as an electron microscope, with a crowded schedule for both the operator and the instrument. The following papers point out what has been accomplished, often using clonal cultures as tools. The reader will see that there is more to learn in each group and will find suggestions for research directions. It is only by growing selected species in the laboratory and taking the results back to the field that we can answer questions about the survival strategies of the algae.

References

Carr, N. G. (1979). Differentiation in filamentous Cyanobacteria. In *Developmental Biology of Procaryotes*, ed. J. H. Parish, pp. 167–201. Oxford: Blackwell Scientific Publications.

Carr, N. G., & Whitton, B. A., eds. (1982). *The Biology of Cyanobacteria.* Oxford: Blackwell Scientific Publications.

Chapman, D. V., Dodge, J. D., & Heaney, S. I. (1982). Cyst formation in the freshwater dinoflagellate *Ceratium hirundinella* (Dinophyceae). *Journal of Phycology*, **18**, 121–129.

Fryxell, G. A., Doucette, G. J., & Hubbard, G. F. (1981). The genus *Thalassiosira:* the bipolar diatom *T. antarctica* Comber. *Botanica Marina*, 24, 321–335.

Garrison, D. L. (1980). *Studies of Coastal Phytoplankton Populations in Monterey Bay, California.* Ph.D. dissertation, University of California, Santa Cruz.

– (1981). Monterey Bay phytoplankton. II. Resting spore cycles in coastal diatom populations. *Journal of Plankton Research*, 3, 137–156.

Nichols, J. M., & Carr, N. G. (1977). Akinetes of Cyanobacteria. In *Spores VII*, ed. G. Chambliss & J. C. Vary, pp. 335–343. American Society for Microbiology, Madison, Wisc.

Schmid, A. M. (1979). Influence of environmental factors on the development of the valve in diatoms. *Protoplasma*, **99**, 99–115.

Stanier, R. Y., & Cohen-Dazire, G. (1977). Phototrophic procaryotes: the Cyanobacteria. *Annual Review of Microbiology*, **31**, 225–274.

Wolk, C. P. (1979). Intercellular interactions and pattern formation in filamentous Cyanobacteria. In *Determinants of Spacial Organization*, eds. S. Subtelny & I. R. Konigsberg, pp. 247–266. New York: Academic Press.

1

The roles of resting spores and akinetes in chlorophyte survival

ANNETTE W. COLEMAN
Division of Biological and Medical Sciences
Brown University
Providence, RI 02912

The majority of the green algae are freshwater or terrestrial organisms. They are subject to the same environmental stresses as their marine cousins: competition for nutrients, escape from poisons and predators, and fluctuations in light and temperature. The overriding problem in their survival and dissemination, however, is water loss. It is not surprising, then, that specialized thick-walled cells of one sort or another occur, not just in the majority, but in the preponderance of the freshwater genera. Among marine forms such cells are almost unknown. In fact, among the predominantly marine orders of Chlorophytes, *Dichotomosiphon* is the only genus known to form akinetes, and it is an inhabitant of freshwater. This chapter concentrates on the thick-walled cell, its nature, and its roles in both short-term and long-term survival of freshwater Chlorophyta.

1.1 Thick-walled cell = resting cell = resistant cell

In classical terminology there are two major types of specialized thick-walled cells among the algae. One, the *akinete*, is a modified vegetative cell in which the wall continues to thicken far more than in metabolically active cells and may also incorporate additional kinds of wall materials. The point of emphasis is that the original cell wall is incorporated into and forms the basis of the final thick wall (Fritsch, 1945). Classic examples are the akinetes of *Pithophora, Zygnema* (Fig. 1), *Tetraspora, Pediastrum,* and *Spongiochloris*.

The second kind of thick-walled cell is represented by the *hypnospore* and *hypnozygote*, both being protoplasts that have separated from their parental wall and participated in some further activity. This may only be formation of a sedentary zoospore or it may involve gametogenesis and

Supported by NSF Research Grants DEB 76-82919, PCM 76-80784, and PCM 79-23054

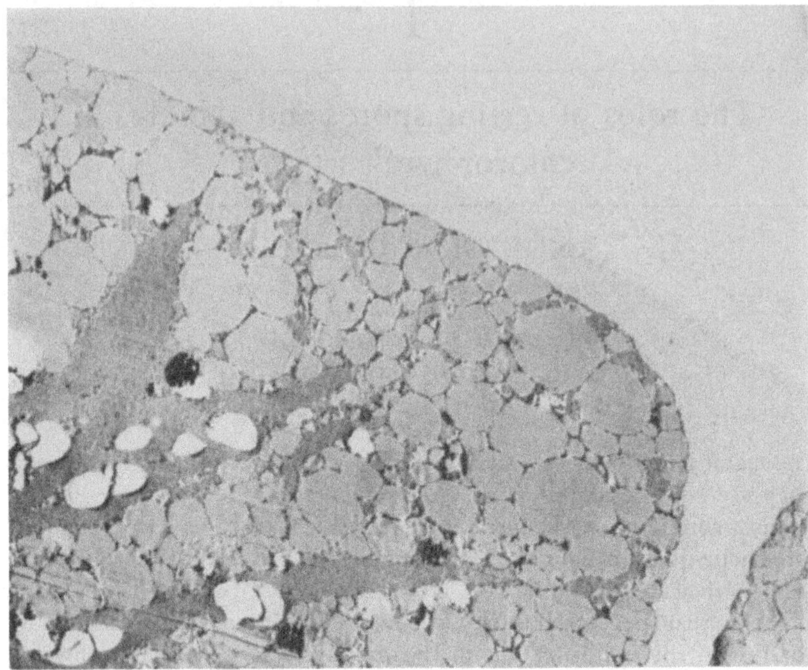

Fig. 1. Electron micrograph of a portion of an akinete of *Zygnema* showing thickened wall and lipid accumulation in the cell interior. Plastid lamellae are no longer distinct. From McLean and Pessoney (1971).

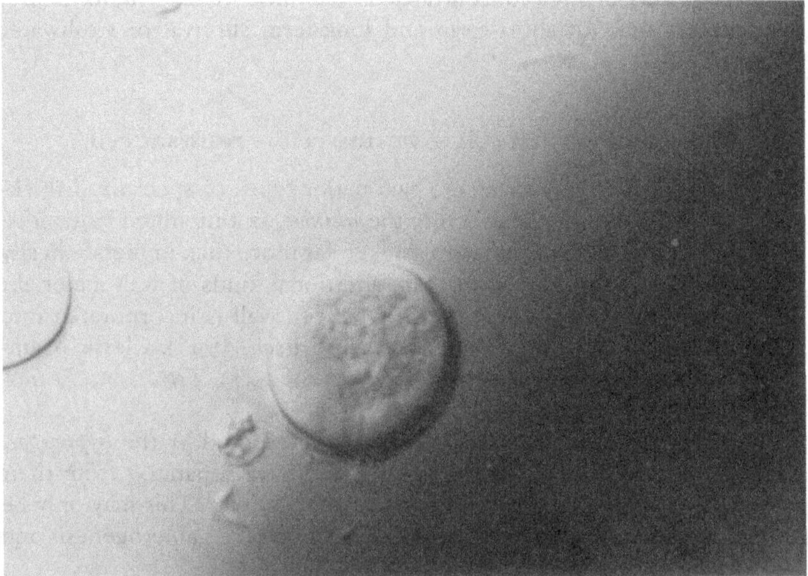

Fig. 2. Living zygospore of *Pandorina* as seen in optical section with Nomarski optics. Large homogeneous-appearing region is lipoidal material.

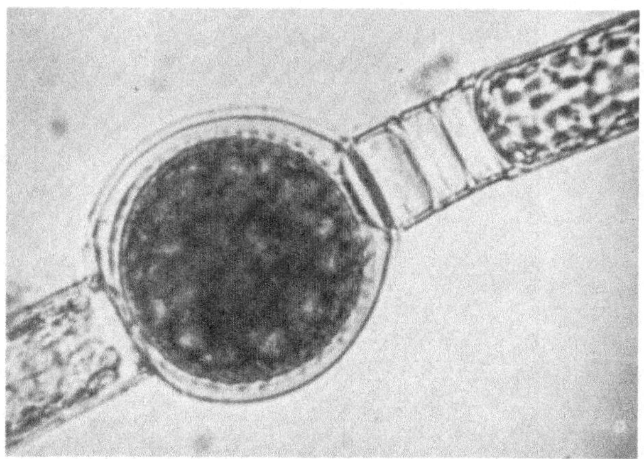

Fig. 3. Light micrograph of a living *Oedogonium* zygote. From Hoffman (1965).

fusion of gametes; these cells then secrete a new, very heavy and usually specially decorated wall. Examples of this behavior include the hypnospores of *Chlorococcum hypnosporum* Starr and *Ulothrix fimbriata* Bold, and the thick-walled zygotes of the Volvocales (Fig. 2), Oedogoniales (Fig. 3), Zygnematales (Figs. 4, 5), and Chlorococcales.

More difficult to define and identify are those usually small, relatively thick-walled vegetative cells imbedded in a heavy gelatinous matrix that must contribute heavily to the survival of their species. In many cases neither specialized akinetes nor hypnospores are known for the species, and yet viable cells can be isolated from air currents or from dried material. Mattox (1971) describes relatively unspecialized cells in *Klebsormidium flaccidum* (Kützing) Silva, Mattox, et Blackwell that resist temperatures up to 100°C, at least briefly, and remain viable after air drying for more than 2 years. Proctor (1966) found that vegetative cells of some desmids could survive passage through the alimentary canal of waterfowl. These are exceptional cases, however, for the vast majority of vegetative cells are far more susceptible to environmental stress than are the specialized thick-walled cells in their life cycles.

The records of akinete and hypnozygote survival are very impressive, but limited so far, primarily by lack of older material to examine. Hoshaw's collection of Zygnematacean zygotes is now more than 20 years old and still viable, as are Hoffman's stocks of *Oedogonium foveolatum* Wittrock (personal communication). Coleman's *Pandorina* zygote collection germinates abundantly after 24 years. Other examples of long-lived zygospores and akinetes stored in the laboratory and remaining viable for periods of years are cited in Coleman (1975). Clear-cut records for field survival are much more difficult to obtain, but Pessoney (1968) suggests that akinetes are adequate for survival capacity in *Zygnema*, whereas zygospores, rather than akinetes, have greater survival capacity

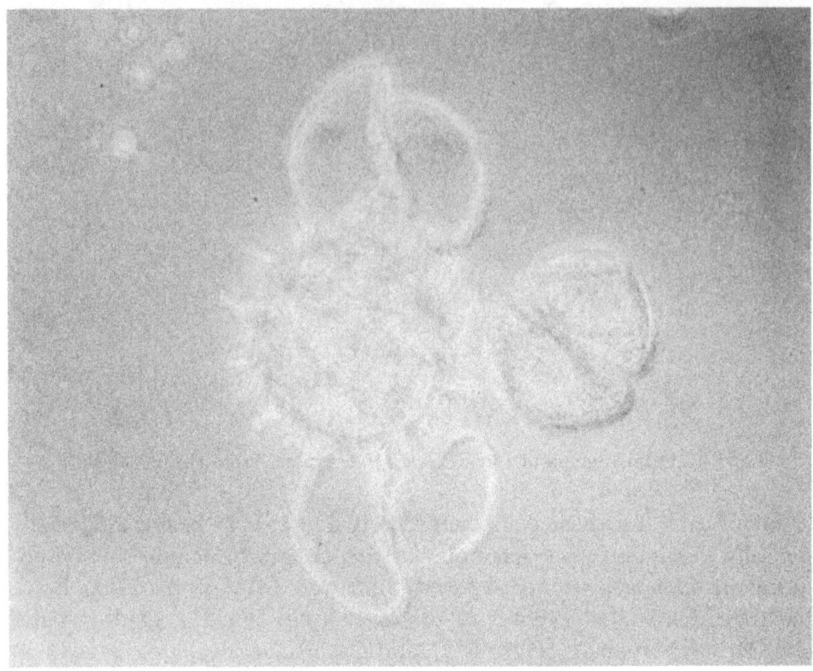

Fig. 4. Zygote, empty vegetative cell walls, and one vegetative cell of *Cosmarium*.

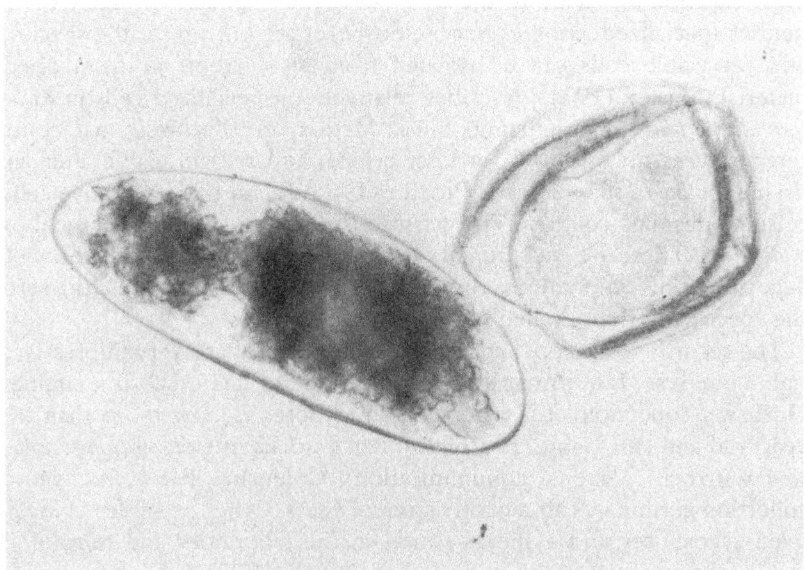

Fig. 5. Germling cell with empty zygospore wall of *Sirogonium*, illustrating the two wall layers left behind at germination. From Hoshaw (1968).

in *Spirogyra*. In a direct study, Lembi et al. (1980) germinated *Pithophora* akinetes cored from a depth of 15 cm below an Indiana lake, a depth that suggests a deposition time of 2–3 years. These reports help to confirm the commonly held presumptions concerning the efficacy of thick-walled spores but also indicate the paucity of direct experimental work in the field. Coring soil samples from recently created lakes, burying identifiable spores or vegetative cells in containers open to the subterranean environment, and even inoculation of single clones or compatible pairs of clones identifiable by mating type into experimental ponds for later recovery (Coleman, 1973) are all possible experimental approaches that would aid our understanding of the importance of thick-walled cells to population survival.

1.2 Dispersal of algae

The groups in which "vegetative cells" contribute heavily to survival under extreme conditions may be more obvious from an examination of the sources of colonizing algae. Over the short term of just a season or a few years, survival is a question of invading new habitats and establishing and maintaining the population from year to year. Geological and climatological events, not to speak of human activities, constantly alter the surface of the earth, destroying old niches as well as creating new ones. No open body of freshwater remains devoid of algae for much longer than the first gust of wind. Algae lurking in the soil, brought in by flood waters or conveyed by animals, including birds, insects, and even raccoons (Bassett, 1963), rapidly contribute to a diverse population.

Although no one has succeeded yet in identifying the precise sources of inocula for a body of water in nature, there has been extensive sampling of viable algae found floating in the atmosphere, on the presumption that atmospheric disturbances, particularly catastrophic ones, could provide adequate inocula of either soil or planktonic forms. Results of three of these studies (Brown et al., 1964; Schlichting, 1964; Brown, 1971) are tabulated in columns A, B, and C of Table 1, along with two results of soil samplings (Holm-Hansen, 1964; MacEntee et al., 1972) in columns E and G, a list of the phytoplankton genera present in more than 10% of water samples taken in the eastern United States (Taylor et al., 1979) in column F, and in column D, a composite listing of those Chlorophyta most tolerant of polluted waters, genera gleaned from more than 100 reports (Palmer, 1969). The genera are listed by orders according to the taxonomic treatment of Bourrelly (1966). Handling such lists by generic name alone begs many questions concerning species diversity and variation, but there is no alternative at present. Similarly, one must recognize that each author who cultivated samples used but a single medium and set of culture conditions, which are predominantly biased

Table 1. *Aerial, terrestrial, and aquatic sampling of chlorophytes*

	A	B	C		D	E	F	G
Chaetophorales								
				Stigeoclonium	+			
Cladophorales								
				Cladophora	+			
Chlorococcales								
				Actinastrum	+		+	
			+*	*Ankistrodesmus*	+	+		
		+	+*	*Bracteacoccus*-like				+*
				Characium				+
	+*	+	+*	*Chlorella*	+	+		+
	+*	+	+*	*Chlorococcum*	+	+		+*
	+			*Coelastrum*	+		+	
				Crucigenia	+		+	
	+			*Dictyochloris*				
				Dictyococcus				+
				Dictyosphaerium	+		+	
				Golenkinia	+		+	
				Kentrosphaera		+		
				Kirchneriella			+	
				Lagerheimia			+	
				Micractinium	+			
	+*	+	+	*Neochloris*-like				+*
				Neospongiococcum				+*
	+			*Oocystis*	+		+	+*
	+			*Palmellococcus*				
				Pediastrum	+		+	
	+			*Planktosphaeria*				+
	+		+	*Protosiphon*				+*
	+			*Radiococcus*				
	+			*Radiosphaera*				+*
	+		+	*Scenedesmus*	+		+	
				Schroederia			+	
				Selenastrum	+			
	+		+	*Spongiochloris*-like				+*
	+			*Spongiococcum*				
			+*	*Tetraedron*	+		+	+
	+*			*Trebouxia*				+
			+	*Treubaria*-like			+	
	+			*Westella*				
Chlorosarcinales								
	+			*Borodinella and Borodinellopsis*				+
	+			*Chlorosarcina*		+		
	+*			*Chlorosarcinopsis*				+*
	+			*Chlorosphaeropsis*				
				Coccomyxa		+		
	+			*Friedmannia*				
	+*		+	*Nannochloris*				
	+			*Ourococcus*				

	A	B	C		D	E	F	G
	+	+		*Tetracystis–Pseudotetracystis*				+*
Oedognoiales								
			+	*Oedogonium*				
Siphonocladales								
			+	*Rhizoclonium*				
Tetrasporales								
			+	*Asterococcus*-like				
				Gloeococcus				+
			+	*Gloeocystis*				+*
	+			*Hormotilopsis*				
	+		+	*Palmella*-like				+*
				Sphaerellocystis				+
			+	*Sphaerocystis*				
	+			*Tetraspora*				
Ultrichales								
				Binuclearia		+		
	+		+	*Klebsormidium*				+*
			+	*Microspora*				
	+			*Pleurastrum*				
	+*		+*	*Pleurococcus–Protococcus*				+
	+			*Pseudulvella*-like				
				Raphidonema		+		
	+			*Stichococcus*		+		+
	+		+	*Ulothrix*	+		+	
Ulvales								
			+	*Prasiola*-like				
Volvocales								
	+	+	+	*Chlamydomonas*	+	+	+	+*
				Chlorogonium	+		+	
				Eudorina	+			
				Gonium	+			+
				Pandorina	+		+	
			+	*Pleodorina*				
				Pyrobotrys	+			
				Spondylomorum	+			
Zygnematales								
				Closterium	+		+	
	+			*Cosmarium*	+		+	
	+			*Cylindrocystis*				
				Euastrum			+	
				Penium				+
	+			*Roya*				
				Spirogyra	+			
				Staurastrum			+	

Note: Air samples: A, B, and C. Water samples: D and F. Soil samples: E and G. *Sources:* A, Brown et al. (1964); B, Brown (1971); C, Schlichting (1964); D, Palmer (1969); E, Holm-Hansen (1964); F, Taylor et al. (1979); G, MacEntee et al. (1972). Asterisks mark genera present in more than five samples.

toward the cultivation of common green algae. Nevertheless, the table offers some interesting comparisons.

The five genera of airborne algae in column B are all also represented in column A and, with one exception, in column C, suggesting that air sampling is remarkably repeatable, and that Michigan, Texas, and Hawaii samples are not particularly different. Half the genera in column C are also listed in column A. Even more striking is the fact that only 8 of the 28 algae found in dry forest soils (column G) are not listed among the air samplings. This suggests that viable soil algal cells are continuously introduced into the air. By contrast, 13 of the 20 most common components of the phytoplankton are not mentioned among either the airborne or the dry soil algae; for most of these, one might merely assume that their numbers were not sufficient to be detected, that is, they exhibit no unique characteristics.

Among the seven most common phytoplankton genera also trapped in air samples, four are listed only among the airborne algae, and the other three, *Oocystis, Tetraedron,* and *Chlamydomonas*, are common to air, soil, and water samples. In view of the number of species in the last-named genus, and their frequent production of heavy-walled resting cells, it comes as no surprise to find *Chlamydomonas* among the most commonly sampled algae. However, the tolerance of *Oocystis* and *Tetraedron* to desiccation during air transport must result, at least in part, from their heavy gelatinous sheaths.

Finally, all but six of the common phytoplankton genera (column F) are also listed among the 27 genera tolerant of polluted waters (column D). Perhaps the simplest interpretation would be that the physiological variation within genera, if not species, may be so great as to make pointless the search for genera that can serve as an index to pollution.

More to the point of our inquiry, only 2 of the 10 most common atmospheric algae are known to have heavy-walled cells in their life cycles, and only 6 out of the 14 most common soil forms. One must conclude that, at least for such small-celled, matrix-imbedded genera, the classic hypnospores and akinetes are not required for short-term survival in stressful situations. In this context, such a conclusion seems valid even if thick-walled cell stages exist of which we are not yet aware (Trainor & Burg, 1965), for they must be relatively rare. Thus, the most frequently transported organisms are not those with unique thick-walled akinetes or spores; they are merely the most abundant algae in soil and water. For these small-celled, gelatinous populations, abundance in numbers may play the same role in dissemination and maintenance of the species that akinetes and hypnospores play for the less-gelatinous forms, or for those algae with larger cells, particularly those with central vacuoles. Without exception, the genera listed in Table 1 that have vegetative cells over 15 μm in any dimension are also known to be able to form specialized thick-walled cells.

1.3 Stable populations of algae

It might be assumed, from the many genera in Table 1, that the algal population of a continent is in a continual whirl, constantly being depleted at a site and being replenished by new immigrants. For several groups, however, there is evidence to suggest that this is not so. Many investigators have their favorite *Spirogyra* pond or some constant yearly source of a particular algal form. That these may represent the same genetic population is clear from the report by Proctor (1975) of recollecting the same unusual *Chara* from an isolated pond in South America that had been described decades earlier, from Biebel's (1973) report on the extremely rare species *Mesotaenium dodekahedron* Geitler, and from the recapture of identical mating types of *Pandorina morum* in the same ponds 18 and 19 years after the original study had been made (Coleman, 1977). This latter study and the one by Stein and McCauley (1976) show that even rather small ponds may have more than one population of a single species, a fact that would not be recognized on morphological grounds alone. Additional identifying characters are necessary to distinguish genetically isolated populations of the same species. Thus, we find that opportunities for immigration are frequent but also that unique and identifiable populations can inhabit the same pond for decades. There seems no basis on which to decide whether the most common algae are good migrators or good adaptors.

Among the orders represented in Table 1, only the Volvocales are outstandingly common in polluted environments, although relatively poorly represented in aerial samples, dry land samples, or even the plankton in general. They also are the only nonfilamentous order in which the majority of genera are known to have hypnospores, hypnozygotes, or akinetes. Particularly here, the thick-walled cell may play a major role in the spread and maintenance of populations. In fact, there are examples of successful laboratory intercrosses between organisms from geographically isolated sites for six species of this order (cited in Coleman, 1977). Although we know now that many morphological species are not cosmopolitan in their hybridization potential, some genetically compatible subpopulations of species may approach being worldwide in distribution.

1.4 Thick-walled cell formation as a developmental pathway

It is well to emphasize that the developmental pathway leading to the production of thick-walled cells in Chlorophytes is not primarily reproductive, in the sense of producing many new organisms. The thick-walled cells, akinetes, hypnospores, or hypnozygotes, are all formed in response to conditions of adequate photosynthesis but limiting nitrogen. Experimental manipulation of sporulation has been documented strik-

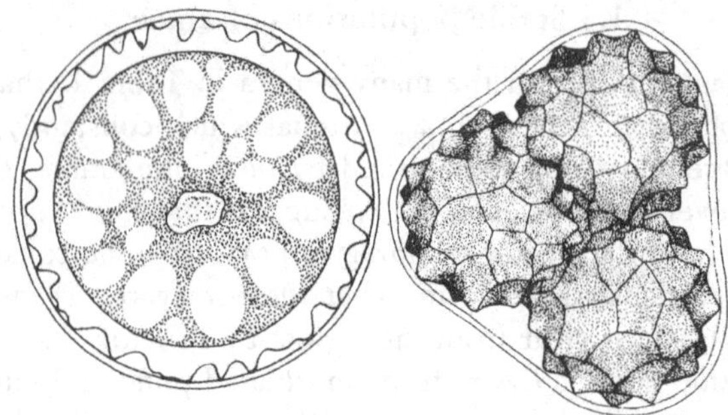

Fig. 6. Group of hypnospores of *Chlorococcum hypnosporum* and median optical section of such a hypnospore. From Starr (1955a).

ingly for akinetes of *Spongiochloris* by McLean (1968), for akinetes and hypnozygotes of Zygnematales by Hoshaw (1968), and for the hypnozygotes of *Chlamydomonas* by Sager and Granick (1954). Similarly, drying of algal mats promotes akinete formation in *Pithophora* (Lembi et al., 1980). Reviews and further citations of methods of inducing production and germination of resistant cells can be found in Coleman (1962), Erben (1962), and Dring (1974). Sager and Granick (1954) pointed out that the same stimulus is recognized by the spore-forming bacteria and that in both cases, resistant spores form just as population numbers peak. Unfortunately, the more specific biochemical nature of the trigger mechanism eludes us, so that we continue to describe the physiological manipulations in terms similar to those used by Klebs (1896) in his classic description of research on algal reproduction.

A factor that has lent confusion to investigations of the physiology of resting cell formation may be the concentration of research on organisms like *Chlamydomonas* that combine sexual reproduction with resting cell formation, the result being the hypnozygote. This type of life cycle, which is the most common among the freshwater Chlorophyta, can be contrasted with that of *Cladophora* or *Ulothrix*: These also make thick-walled cells (akinetes) as the environment becomes less hospitable for growth, but their reproductive activity occurs at a physiologically quite different portion of the life cycle, at the time when signals of a fresh growth period appear in their environment or when light, but not nitrogen, may be limiting.

The notion that resting cell formation represents a unique and separable developmental pathway can be inferred not only from the variations in life cycles of different genera but also experimentally. In *Chlamydomonas* (Ebersold, 1967), diploid cells can be obtained by nutritional selection methods. The cells are the product of normal gamete pairing and fusion but then omit the resting spore stage and its usual outcome, meiosis at zygote germination. Instead, such cells remain vegetative and

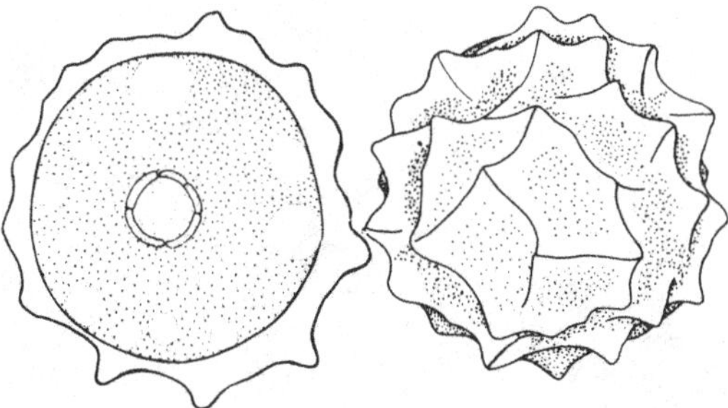

Fig. 7. Surface view and median optical section of *Chlorococcum echinozygotum* zygospore formed after gamete fusion. From Starr (1955a).

engender normal but diploid *Chlamydomonas* cultures. Thus, the events of gametogenesis and fusion have occurred without simultaneously including resting spore formation. The situation in which resting cell formation takes place without the expected accompaniment of gametogenesis and cell fusion is illustrated (Figs. 6, 7) by comparing *C. hypnosporum* with *Chlorococcum echinozygotum* Starr (Starr, 1955a). In the former, potential gametes/zoospores form hypnospores, whereas in the latter, potential gametes/zoospores pair, fuse, and form hypnozygotes.

1.5 Cell walls of hypnospores, hypnozygotes, and akinetes

If resting spore formation represents a distinct line of cell differentiation, are there characteristics common to all three forms, the akinete, the hypnospore, and the hypnozygote? Despite variations in gross appearance, there may be more similarities than have been realized. The thickened wall fundamentally has three parts, although the outermost of these is often missed under the conditions used for fixation for electron microscopy. This outermost layer is most readily demonstrated in India ink mounts of living cells. To some extent its components continually dissolve into the culture medium, but when motile zygotes settle, this material provides the glue that fastens developing zygotes to each other and to surrounding surfaces (Cavalier-Smith, 1976). In electron micrographs, where this structure has been well preserved (e.g., Fulton, 1978), the fibrous material in this layer is aligned normal to the cell surface.

Just within this material lies a tripartite layer that is visible as a bright line in the light microscope and as an alternation of dark–light–dark lines when normal sections are seen in electron micrographs ("central lamina" of Fig. 8). In position, the layer corresponds to the primary zygote wall (Starr, 1955b) of light microscopy and to the insoluble trilaminar sheath of Staehelin and Pickett-Heaps (1975). Considerably more structure is found in this layer, albeit it is as yet poorly understood. Its width varies from about 0.05 to 0.15 μm, depending upon the group of organisms,

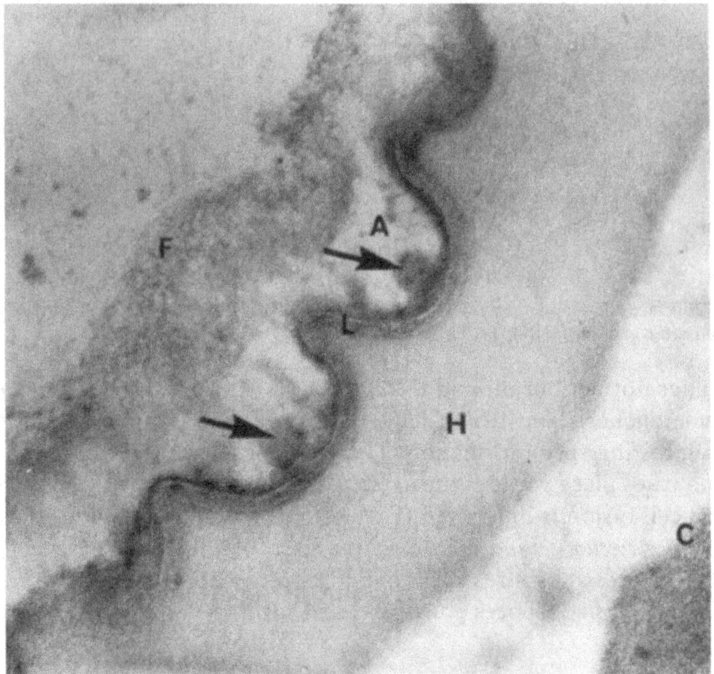

Fig. 8. Electron micrograph of transverse section of mature zygospore wall of *Chlamydomonas reinhardtii.* Cell surface is at C and tripartite wall layer at L. Other designations indicate substucture of the wall. From Cavalier-Smith (1976).

and it is a remarkably constant component of cell walls in many genera, both in the vegetative wall and that thickened for resting spores. The filamentous orders with larger cells, the latter each containing a central vacuole, may not have such a complex layer at this position (Hoshaw, 1980); instead they appear to have outer- and innermost layers of cellulose and median layers, the most variable in thickness, of unknown materials (Parker, 1964).

Internal to the tripartite layer lies a zone of diverse appearance and thickness, but one where filamentous components are aligned parallel to the cell surface. According to Starr (1955b), who watched the deposition of the material in maturing zygotes of *Cosmarium botrytis* Meneghini, cells only become resistant to drying after this layer, the secondary zygote wall, has appeared. Within it are formed the various ornamentations of thick-walled cells.

It is not yet clear whether the walls of the resistant cells differ qualitatively from the walls of their respective vegetative cells, although at least two reports, both concerning *Chlamydomonas reinhardtii* Dangeard, suggested that this is so. Minami and Goodenough (1978) reported finding new polypeptides and glycopolypeptides beginning at about the time of nuclear fusion in the zygote. Catt (1979) compared the wall composition in vegetative cells and zygotes of *C. reinhardtii.* He found only 5% protein in the zygote wall compared to 34% in the vegetative cell wall. Furthermore, the zygote wall contained almost 50% glucose residues in its carbohydrate analysis, while vegetative cell walls contained only traces of glucose but an abundance of other sugars. Finally, amino acid analysis showed that not only is tryptophan missing, as in vegetative walls, but also histidine, whereas proline and hydroxyproline remained relatively high. These zygote wall polypeptide chains, then, display the collagenlike amino acid composition characteristic of the vegetative wall material but tend even more toward elimination of aromatic amino acids. This perhaps reflects their role of remaining metabolically inert for long periods of time in the presence of bright light and oxygen, which are known to cause photodynamic damage to certain aromatic amino acids (Goosey et al., 1980).

As Catt (1979) noted, nearly 25% of the dry weight of the *Chlamydomonas* zygote wall material remained unaccounted for by carbohydrate or proteinaceous material. Among the other possible materials that have not usually been sought in chemical analysis are a variety of heavy ions, such as iron and manganese (Schulz-Baldes & Lewin, 1975). These were thought by Pringsheim (1953) to be required for the color and thickening of the envelope of *Trachelomonas* and very probably contribute to the rust color of various zygotes in strains where the pigmentation is known to lie in the wall or to be encrusted (Lewin, 1957). Algal wall materials are documented cation-exchange matrices (Mongar & Wassermann, 1952; Button & Hostetter, 1977). Esterified sulfate groups are one obvious source of cation binding sites. These have now been identified chemically in *Pandorina* wall material (Fulton, 1978) and in *Platydorina* walls (Crayton, 1980) and probably deserve more attention among Chlorophyte algae. Early reports of the presence of chitin in green algal walls were not confirmed by the investigation of Parker (1964), who examined only vegetative cells. However, the report of Lembi et al. (1980) on chitin in *Pithophora* akinetes suggests that further work should be done. Finally, in *Chlorella* and *Scenedesmus,* the material that remains after both acid and base extraction of walls is suggested on this basis to be sporopollenin, a carotenoid polymer characteristic of the exine of pollen grains (Brooks & Shaw, 1968; Atkinson et al., 1972; Staehelin & Pickett-Heaps, 1975).

1.6 Hypnospore, hypnozygote, and akinete contents

The impermeability of the light-matured thick wall is impressive, at least in types such as the *Chlamydomonas* zygotes tested by Lewin (1957).

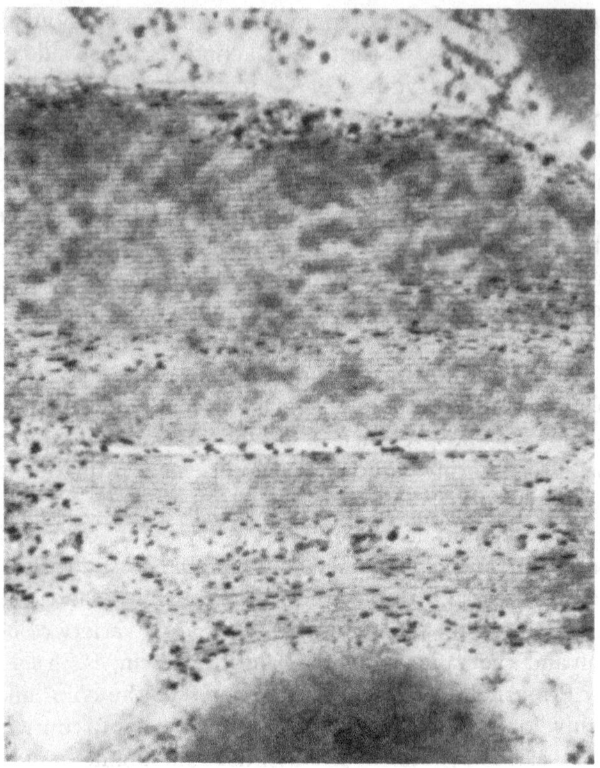

Fig. 9. Electron micrograph of transverse section of a portion of the chloroplast in a mature zygospore of *Chlamydomonas moewusii*. The laminae display an almost crystalline arrangement. From Brown et al. (1968).

These retained their viability after treatment with arsenate, cyanide, 75% sulfuric acid, saturated sodium chloride, and even 12 days in acetone. Lembi et al. (1980) report that *Pithophora* akinetes can survive 12 hr in up to 8 μg/ml copper, an observation that, combined with the vast numbers of akinetes deposited over lake bottoms, augurs poorly for any copper sulfate control system. *Pandorina morum* zygotes survive such concentrations also (Coleman, unpublished observations).

This resistance helps to explain resistant cell longevity but challenges us to explain how organelles and cytoplasm can be packaged in such a way as to maintain themselves in suspended animation at growth temperatures. The answer is universally assumed to lie in the state of dehydration of the resistant cells' contents. Both photosynthesis and respiration seem to be shut down to a level below detection, and chlorophyll is generally not evident. Cavalier-Smith (1976) has emphasized the cytoplasmic density seen in electron micrographs of *Chlamydomonas* zygotes. Other rare micrographs suggest a remarkably tight packing of plastid membranes (Fig. 9), or an extensive degeneration of the entire set of

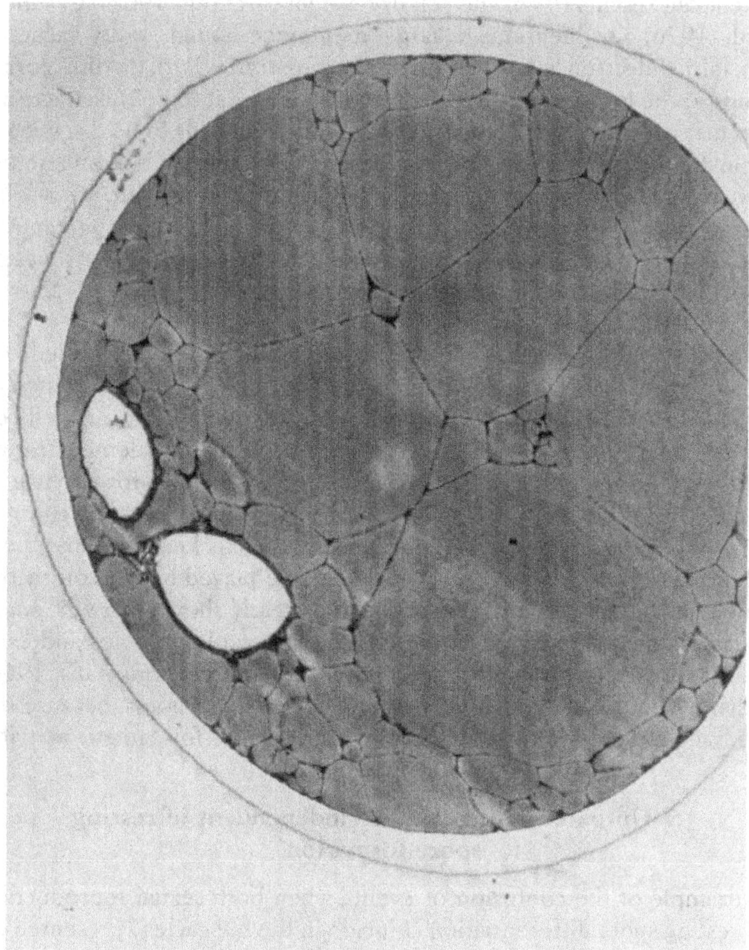

Fig. 10. Electron micrograph of section through the heavy-walled, lipid-filled akinete of *Spongiochloris*. From McLean (1968).

plastid lamellae (Brown et al., 1968; McLean, 1968; Pickett-Heaps, 1975). It is conceivable that an organelle comparable to a proplastid can be found in heavy-walled spores; proplastids have now been reported in several *Chlamydomonas* strains grown for long periods in the dark (Friedberg et al., 1970). Very little is known of mitochondria, but Hommersand and Thimann (1965) have reported major changes in the functional cytochromes as *C. reinhardtii* zygotes begin germination, changes that occur before any photosynthesis is detectable.

Carotenoid coloration is distinctive in many thick-walled cells, particularly akinetes, and storage product accumulation, starch and lipid, is standard (Fig. 10). In *Spongiochloris* akinetes, the red-orange carotenoid content can reach 41% of the cell dry weight (McLean, 1967). Cavalier-Smith (1976) also mentions a "large, membrane-bound empty vacuole" seen in his electron micrographs of *C. reinhardtii.* Perhaps this corresponds to the location of the cap of lipoidal material (Fig. 2) characteristic of many mature Volvocalean zygotes. Much remains to be discovered within the thick wall, but as Pickett-Heaps (1975) says: "Electron microscopy of maturing zygospores becomes increasingly difficult. The cells' accumulation of starch, lipid droplets, etc., makes processing for electron microscopy less and less rewarding, and the cytoplasm finally becomes so dense that ultrastructural details are often totally obscured." This is clearly a challenge for future investigation.

An organelle of particular interest to understanding thick-walled cell formation and germination is the contractile vacuole. If the reentry of water is the first stimulus to germination of the highly concentrated cell contents, and the thick wall appears to crack instead of demonstrating elasticity at germination, one of the earliest prerequisites for survival of the contents of the thick-walled cell must be a system of osmotic control. In *Chlamydomonas* zygotes, the contractile vacuole is known to disappear with maturation (Brown et al., 1968). The role played by the contractile vacuole in handling the osmotic problem is clearly shown by two *Chlamydomonas* mutants, each appearing to lack contractile vacuoles and each requiring media of elevated tonicity in order to survive (Gowans, 1960; Guillard, 1960). Such mutants could not even form zygotes, because the cells lysed in media with sufficiently low osmolarity for gametic activity.

1.7 Uniparental inheritance independent of resting spore formation

An example of the confusion of events when both sexual reproduction and resting spore differentiation coincide in the life cycle is presented by the many observations related to plastid inheritance. Cytological evidence for the fusion of gamete plastids at about the time of nuclear fusion has been presented for zygotes of *Chlamydomonas moewusii* Gerloff (Brown et al., 1968) and *C. reinhardtii* (Cavalier-Smith, 1970). At germination of *C. moewussi* zygotes, both plastid pyrenoids are said to survive (Brown et al., 1968), whereas in *C. reinhardtii* (Cavalier-Smith, 1976) but a single pyrenoid survives, and there are indications that some plastid degeneration occurs. The latter point is unclear because it was not reported whether these were zygotes that gave 90%–100% germination. Genetic evidence (Sager, 1954; Gowans, 1960) on the other hand, indicates that the genetic elements of only one of the gametic plastids survive among the zygote progeny.

Various reports for Zygnematalean and Ulvalean algae (summarized in Lewin, 1976), citing morphological and even radioautographic data, suggest that the plastid(s) donated by one gamete is degraded in the zygote while that of the other survives. That this happens with *Ulva*, where the zygote is not a thick-walled resting cell, may certify that uniparental inheritance is entirely associated with sexual reproduction and that the reduction of plastid structure concomitant with resting spore formation has only confused our understanding.

1.8 Effect of resting cells on diversity and selection

Survival of species is a problem not of just a season or a few years but of time beyond our everyday comprehension. In theoretical population biology, it is accepted that any population with variation in its genetic pool has the potential advantage in long-term survival. The advantage lies not just in immediately making new and selectively advantageous gene combinations but also in having a reservoir of genetic variation to cope with alterations in the environment. The problem this poses, theoretically, for freshwater Chlorophytes is that almost all are haploid in their vegetative state, and consequently there is no heterozygous reservoir for variant alleles.

As in all organisms, the ultimate source of variation is mutation, but with haploid organisms, selection is severe. For example, if a new allele arose that conferred on the phenotype of its organism only half the fitness of its competitor, and one started in a new pond with a unicell of each genotype, the frequency of the less fit allele would be expected to be less than 1 in 100 after only six mitotic generations. Conversely, of course, selection for an allele conferring some favorable phenotype would progress just as rapidly. The point is that variety would seem to be eliminated rapidly. Various solutions to this perplexity have been suggested (e.g., Feldman, 1971). However, an obvious possibility, given the remarkable survival ability of thick-walled cells, is that these may serve as a major reservoir of diversity for a species. Each time buried cells are brought to water and light by flood or physical disturbance, there is a potential for a past generation to reappear. This is a problem which formal population genetics has not yet dealt with, but combined with the widespread occurrence of sexuality among these algae, it provides a rich field for speculation. This constant replay of past generations, crossing back into parts of the gene pool, may help to circumscribe the association of genetic elements characteristic of a recognizable morphological species.

For those organisms that unite the roles of zygote and resting cell, there are two obvious consequences. The genetic material put into long storage represents the greatest genetic diversity and potential for new genotypes available in the life history. The second consequence is that there is strong selection for sexual reproduction, for the only way to produce resistant cells is by mating. An evaluation of how this affects population structure within a species is a challenging task for the future.

To summarize, many small-celled green algae do not make elaborate thick-walled cells but survive short periods of air exposure without difficulty in the vegetative condition. Large-celled (≥15 μm) green algae do not. It is these algae that elaborate akinetes, hypnospores, and hypnozygotes. Details of the structural components of specialized thick walls are virtually unknown. Likewise, the organization of the internal contents of such walls, presumably partially desiccated and reduced in function, is not understood. Finally, truly long-lived resting cells can serve not only to spread the species but also to sequester genotypes from selection over periods of years while the local environment may fluctuate. The reentry, upon resting cell germination, of such genotypes into the active population is a factor of unmeasured importance to the population biology of freshwater chlorophytes.

References

Atkinson, A. S., Gunning, B. E. S., & John, P. C. L. (1972). Sporopollenin in the cell wall of *Chlorella* and other algae: ultrastructure, chemistry, and incorporation of ^{14}C-acetate, studied in synchronous cultures. *Planta,* **107**, 1–32.

Bassett, M., Jr. (1963). The passive dispersal of small aquatic organisms and their colonization of isolated bodies of water. *Ecological Monographs,* **33**, 161–185.

Biebel, P. (1973). Morphology and life cycles of saccoderm desmids in culture. *Beihefte zur Nova Hedwigia,* **42**, 39–57.

Bourelly, P. (1966). *Les algues d'eau douce. Initiation à la systematique. I. Les algues vertes.* N. Boubée, Paris.

Brooks, J., & Shaw, G. (1968). Chemical structure of the exine of pollen walls and a new function for carotenoids in nature. *Nature (London),* **219**, 523–533.

Brown, R. M., Jr. (1971). Studies of Hawaiian algae. I. The atmospheric dispersal of algae and fern spores across the island of Oahu, Hawaii. In *Contributions in Phycology*, ed. B. C. Parker & R. M. Brown Jr., pp. 175–188. Lawrence, Kans.: Allen Press.

Brown, R. M., Jr., Johnson, C., & Bold, H. C. (1968). Electron and phase-contrast microscopy of sexual reproduction in *Chlamydomonas moewusii. Journal of Phycology,* **4**, 100–120.

Brown, R. M., Jr., Larson, D. A., & Bold, H. C. (1964). Airborne algae: their abundance and heterogeneity. *Science,* **143**, 583–585.

Button, K. S., & Hostetter, H. P. (1977). Copper sorption and release by *Cyclotella meneghiniana* (Bacillariophyceae) and *Chlamydomonas reinhardtii* (Chlorophyceae). *Journal of Phycology,* **13**, 198–202.

Catt, J. W. (1979). The isolation and chemical composition of the zygospore cell wall of *Chlamydomonas reinhardtii. Plant Science Letters,* **15**, 69–74.

Cavalier-Smith, T. (1970). Electron microscopic evidence for chloroplast fusion in zygotes of *Chlamydomonas reinhardtii. Nature (London),* **228**, 333–335.

– (1976). Electron microscopy of zygospore formation in *Chlamydomonas reinhardtii. Protoplasma,* **87**, 297–315.

Coleman, A. W. (1962). Sexuality. In *Physiology and Biochemistry of Algae*, ed. R. A. Lewin, pp. 711–729. New York: Academic Press.

– (1973). The use of genetically marked algal strains to study their natural history. *Journal of Phycology*, 9 (Suppl.), 12.

– (1975). Long-term maintenance of fertile algal clones: experience with *Pandorina* (Chlorophyceae). *Journal of Phycology*, **11**, 282–286.

– (1977). Sexual and genetic isolation in the cosmopolitan algal species *Pandorina morum*. *American Journal of Botany*, **64**, 361–368.

Crayton, M. A. (1980). Presence of a sulfated polysaccharide in the extracellular matrix of *Platydorina caudata* (Volvocales, Chlorophyta). *Journal of Phycology*, **16**, 80–87.

Dring, M. J. (1974). Reproduction. In *Botanical Monographs*, Vol. 10, *Algal Physiology and Biochemistry*, ed. W. D. P. Stewart, pp. 814–837. Berkeley: University of California Press.

Ebersold, W. T. (1967). *Chlamydomonas reinhardtii:* heterozygous diploid strains. *Science*, **157**, 447–449.

Erben, K. (1962). Sporulation. In *Physiology and Biochemistry of Algae*, ed. R. A. Lewin, pp. 701–710. New York: Academic Press.

Feldman, M. W. (1971). Equilibrium studies of two locus haploid populations with recombination. *Theoretical and Population Biology*, **2**, 299–318.

Friedberg, I., Goldberg, I., & Omad, I. (1970). The structure of the prolamellar body in *Chlamydomonas reinhardtii* and its possible role in the biogenesis of the photosynthetic lamellae. *7th International Congress of Electron Microscopy*, **3**, 189–190.

Fritsch, F. E. (1945). *The Structure and Reproduction of the Algae*, Vol. I, p. 39. Cambridge: Cambridge University Press.

Fulton, A. B. (1978). Colonial development in *Pandorina morum*. I. Structure and composition of the extracellular matrix. *Developmental Biology*, **64**, 224–235.

Goosey, J. D., Zigler, J. S., Jr., & Kinoshita, J. H. (1980). Cross-linking of lens crystallins in a photodynamic system: a process mediated by singlet oxygen. *Science*, **208**, 1278–1280.

Gowans, C. S. (1960). Some genetic investigations of *Chlamydomonas eugametos*. *Zeitschrift für Induktive Abstammungs- und Vererbungs lehre*, **91**, 63–73.

Guillard, R. R. L. (1960). A mutant of *Chlamydomonas moewusii* lacking contractile vacuoles. *Journal of Protozoology*, **7**, 262–268.

Hoffman, L. R. (1965). Cytological studies of *Oedogonium*. I. Oospore germination in *O. foveolatum*. *American Journal of Botany*, **52**, 173–181.

Holm-Hansen, O. (1964). Isolation and culture of terrestrial and freshwater algae of Antarctica. *Phycologia*, **4**, 43–51.

Hommersand, M. H. & Thimann, K. V. (1965). Terminal respiration of vegetative cells and zygospores in *Chlamydomonas reinhardtii*. *Plant Physiology*, **40**, 1220–1227.

Hoshaw, R. W. (1968). Biology of the filamentous conjugating algae. In *Algae, Man and the Environment*, ed. D.F. Jackson. New York: Syracuse University Press.

Hoshaw, R. W. (1980). Systematics of the Zygnemataceae (Chlorophyceae). II. Zygospore-wall structure in *Sirogonium* and a taxonomic proposal. *Journal of Phycology*, **16**, 242–250.

Klebs, G. (1896). *Die Bedingungen der Fortpflanzung bei einigen Algen und Pilzen.* Verlag von Gustav Fisher, Jena.

Lembi, C. A., Pealmutter, N. L., & Spencer, D. F. (1980). *Life Cycle, Ecology, and Management Considerations of the Green Filamentous Alga, Pithophora*, Technical Report No. 130, pp. 1–97. West Lafayette, Ind.: Purdue University Water Resources Center.

Lewin, R. A. (1957). The zygote of *Chlamydomonas moewusii*. *Canadian Journal of Botany*, **35**, 795–804.

– ed. (1976). *Botanical Monographs*, Vol. 12, *The Genetics of Algae*. Berkeley: University of California Press.

MacEntee, F. J., Schreckenberg, G., & Bold, H. C. (1972). Some observations on the distribution of edaphic algae. *Soil Science*, **114**, 171–179.

McLean, R. J. (1967). Desiccation and heat resistance of the green alga *Spongiochloris typica*. *Canadian Journal of Botany*, **45**, 1933–1939.

– (1968). Ultrastructure of *Spongiochloris typica* during senescence. *Journal of Phycology*, **4**, 277–283.

McLean, R. J., & Pessoney, G. F. (1971). Formation and resistance of akinetes of *Zygnema*. In *Contributions in Phycology*, ed. B. C. Parker & R. M. Brown, Jr., pp. 145–152. Lawrence, Kans.: Allen Press.

Mattox, K. R. (1971). Zoosporogenesis and resistant-cell formation in *Hormidium flaccidum*. In *Contributions in Phycology*, ed. B. S. Parker & R. M. Brown, Jr., pp. 137–144. Lawrence, Kans.: Allen Press.

Minami, S. A., & Goodenough, U. W. (1978). Novel glycopolypeptide synthesis induced by gametic cell fusion in *Chlamydomonas reinhardtii*. *Journal of Cell Biology*, **77**, 165–181.

Mongar, I. L., & Wassermann, A. (1952). Absorption of electrolyte by alginate gels without and with cation exchange. *Journal of the American Chemical Society*, **1952**, 492–497.

Palmer, C. M. (1969). A composite rating of algae tolerating organic pollution. *Journal of Phycology*, **5**, 78–82.

Parker, B. C. (1964). The structure and chemical composition of cell walls of three Chlorophycean algae. *Phycologia*, **4**, 63–74.

Pessoney, G. F. (1968). *Field and Laboratory Investigation of Zygnemataceous Algae*. Ph.D. Thesis, The University of Texas, Austin.

Pickett-Heaps, J. D. (1975). *Green Algae*. Sunderland, Maine: Sinauer Association.

Pringsheim, E. G. (1953). Observations on some species of *Trachelomonas* grown in culture. *New Phytologist*, **52**, 238–266.

Proctor, V. W. (1966). Dispersal of desmids by waterbirds. *Phycologia*, **5**, 227–232.

– (1975). The nature of chlorophyte species. *Phycologia*, **14**, 100–103.

Sager, R. (1954). Mendelian and non-Mendelian inheritance of streptomycin resistance in *Chlamydomonas*. *Proceedings of the National Academy of Sciences USA*, **40**, 356–362.

Sager, R., & Granick, S. (1954). Nutritional control of sexuality in *Chlamydomonas reinhardtii*. *Journal of General Physiology*, **37**, 729–742.

Schlichting, H. E., Jr. (1964). Meterological conditions affecting the dispersal of airborne algae and protozoa. *Lloydia*, **27**, 64–78.

Schulz-Baldes, M., & Lewin, R. A. (1975). Manganese encrustation of zygospores of a *Chlamydomonas* (Chlorophyta: Volvocales). *Science*, **188**, 1119–1120.

Staehelin, L. A., & Pickett-Heaps, J. D. (1975). The ultrastructure of *Scenedesmus* (Chlorophyceae). I. Species with the "reticulate" or "warty" type of ornamental layer. *Journal of Phycology*, **11**, 163–185.

Starr, R. C. (1955a). *A Comparative Study of Chlorococcum meneghini*, Indiana University Publications, Science Series No. 20. Bloomington: Indiana University Press.

– (1955b). Zygospore germination in *Cosmarium botrytis var. subtumidium. American Journal of Botany,* **42,** 577–581.

Stein, J. R., & McCauley, M. J. (1976). Sexual compatibility in *Gonium pectorale* (Volvocales: Chlorophyceae) from soil of a single pond. *Canadian Journal of Botany,* **54,** 1121–1130.

Taylor, W. D., Hern, S. C., Williams, L. R., Lambou, V. W., Morris, M. K., & Morris, F. A. (1979). *Phytoplankton Water Quality Relationships in U.S. Lakes. VI. The Common Phytoplankton Genera from Eastern and Southeastern Lakes.* Las Vegas, Nev.: Environmental Monitoring and Support Lab., U.S. Environmental Protection Agency.

Trainor, F. R., & Burg, C. A. (1965). *Scenedesmus obliquus* sexuality. *Science,* **148,** 1094–1095.

2

Survival strategies of chrysophycean flagellates: reproduction and the formation of resistant resting cysts

CRAIG D. SANDGREN

Department of Biology
University of Texas at Arlington
Arlington, TX 76019

Many planktonic microalgae are characterized by having developed a resistant resting stage as a part of their life histories. Cells in the resting stage are comparatively immune to environmental stress and so can serve as a refuge population for recolonization of the plankton following periods of unfavorable environmental conditions. These resting stages may be formed as the result of either sexual or asexual processes; they may be morphologically specialized structures such as resistant cysts or they may involve primarily physiological changes in the vegetative cells themselves resulting in decreased metabolic activity or the production of protective mucilage layers surrounding the cells (palmelloid stage). Each of the algal classes typically represented in marine or freshwater plankton assemblages (Bacillariophyceae, Chlorophyceae, Chrysophyceae, Cryptophyceae, Cyanophyceae, Dinophyceae, Euglenophyceae, Prymnesiophyceae) is characterized by one or several types of resistant resting stages that are often morphologically unique to that class. This information is reviewed in many general reference texts in phycology (Fritsch, 1935; Smith, 1950; Bourrelly, 1968; Bold & Wynne, 1978).

The resting stages are a necessary component of the life history strategies of many phytoplankton. They allow for survival during time periods when the physical or chemical conditions of the planktonic habitat are beyond the physiological tolerances of the vegetative cells. Because the annual variability of freshwater and neritic marine planktonic habitats is

The author would like to express his appreciation to Drs. David J. Hibberd and Larry R. Hoffman for their cooperation in allowing the reproduction of their micrographs here. The culture studies presented were carried out at Friday Harbor Laboratory, University of Washington, Friday Harbor, WA, and at Hobart and William Smith Colleges, Geneva, NY. The support of these institutions is gratefully acknowledged. Dr. John T. Lehman was integrally involved in the field studies that generated the data presented here on phytoplankton population dynamics.

apparently broader than the tolerance ranges of many individual phytoplankton species, these organisms are necessarily observed to be seasonal in occurrence (Margaleff, 1958; Hutchinson, 1967). Such a seasonal growth pattern can only be successful if these algae possess a resistant resting stage that can serve to reestablish the vegetative cell population during the next seasonal cycle. Obviously, an understanding of the environmental and physiological factors influencing the alternation between vegetative and resting phases of the life cycle is of extreme importance for interpreting and eventually predicting the dynamics of natural phytoplankton populations.

Golden algae of the class Chrysophyceae are perhaps some of the most strikingly seasonal species common in freshwater. Species of *Dinobryon, Synura,* or *Mallomonas* are often dominant members of the phytoplankton community during the spring and autumn but are largely absent during the summer months (Ruttner, 1930; Hutchinson, 1967). For these species, a siliceous resting cyst termed the *statospore* is the alternative phase in the life history (Scherffel, 1911; Bourrelly, 1957). The statospore is an endogenously formed, spherical structure containing a single pore plugged with material thought to be silicopectic in composition (Bourrelly, 1963). It is these resistant cysts that serve as a benthic repository for some fraction of the vegetative cell population during periods of unfavorable growth conditions. They eventually serve as the sedimentary "seed" source from which the planktonic habitat is repopulated when environmental conditions will again support the growth of vegetative cells.

The biology of the statospore has been poorly understood despite their importance for the successful perennation of chrysophycean flagellates. There is a wealth of classical and modern literature describing the morphology and development of statospores, but these reports are often sketchy or even contradictory; perhaps this is because they are based on brief light microscopical observations of fortuitous field collections or because they concern many diverse and poorly understood species. There is also a wealth of literature concerning the genetic significance of statospores, but these reports have not led to a clear understanding of the statospore's function in the chrysophycean life history. Both sexual and asexual functions have been ascribed to statospores produced by a single species; in some cases these two types of cysts are morphologically indistinguishable (Bourrelly, 1957; Kristiansen, 1965), whereas in other cases they differ only in size (Skuja, 1950). Sexual statospores have only been clearly documented in relatively rare species, whereas the reproductive significance of the cysts produced by common planktonic genera, such as *Dinobryon, Mallomonas, Synura*, and *Uroglena*, remains contradictory or unknown.

The study of the chrysophycean statospore has recently received renewed interest largely through the application of electron microscopical techniques to the analysis of statospore structure and development. In addition, the publication of recipes for defined-culture media (DY III, MWC) in which chrysophycean species can be readily grown (Lehman, 1976a, 1976b) has greatly enhanced our ability to investigate the production of statospores under defined and reproducible conditions. As a result of these recent innovations and subsequent research, understanding of statospore biology has been greatly increased.

A review of the survival strategies of chrysophycean flagellates naturally focuses on the biology to the statospore. This chapter reviews the established literature concerning the structure, development, and genetic significance of these resistant cysts in the light of recent advances in understanding based on fine structural and defined-culture techniques. The encystment dynamics of natural chrysophyte populations will also be examined in order to document the existing variability and then to investigate the significance of this variability with respect to the total size of the refuge population of cysts ultimately deposited on the lake sediments.

2.1 Development of the statospore

As Scherffel (1911) first proposed, the ability to produce a siliceous cyst with a single pore is a characteristic shared by members of the class Chrysophyceae. Pascher (1924, 1932) asserted that the cyst is always composed of two pieces – the cyst wall itself and the plug that fills the single pore. He then used this concept to compare the statospore to the resting cysts of the Xanthophyceae and the Bacillariophyceae in support of combining these two classes with the Chrysophyceae in the division Chrysophyta.

The sizable literature concerning the process of statospore formation as observed with the light microscope has been reviewed a number of times (Hollande, 1952; Bourrelly, 1957; Meyer, 1965; Sheath et al., 1975; Sandgren, 1978, 1980a). As a result of these light microscopical observations, a generally consistent view of the encystment process was obtained. The siliceous cyst wall was known to develop within an internal membrane system – the cyst membrane (Cienkowski, 1870; Prowazek, 1903; Scherffel, 1911; Doflein, 1923; Haye, 1930; Pascher, 1932; Skuja, 1950) – which is now thought to be homologous to the silica deposition vesicle (SDV) in which the silica valves of diatoms are produced (Drum & Pankratz, 1964; Hibberd, 1977; Sandgren, 1980a). The cytoplasm of the encysting cell was observed to be partitioned into an extracystic and an intracystic fraction by the development of the cyst membrane. The nucleus, plastid(s), chrysolaminarin vacuole, and most of the ground cytoplasm were contained within the cyst membrane, whereas the contractile vacuole and a small amount of cytoplasm were consistently excluded (Cienkowski, 1870; Scherffel, 1911; Doflein, 1923; Geitler, 1935; Bourrelly, 1957). Once the cyst wall was completely formed within the cyst membrane, a plug was observed to be deposited within the cyst pore. The cytoplasm became increasingly dense or shrunken and tightly packed with both chrysolaminarin and lipid storage reserves.

Although the encystment process was observed to be greatly similar in all chrysophyte species observed with the light microscope, a number of significant variations were found among species. A survey of this literature (Sandgren, 1980a) revealed variation with respect to the fate of cytoplasm initially outside of the cyst membrane. Many species conserved the extracystic cytoplasm by drawing it into the cyst through the pore upon completion of the cyst wall (Scherffel, 1911; Haye, 1930), although other species apparently lacked this ability so that the extracystic cytoplasm was sloughed off and lost to the mature cyst (Prowazek, 1903; Doflein, 1923; Meyer, 1965). A second point of interspecific variation revolved around a disagreement in early reports as to whether the mature cyst contained the same number of plastids and nuclei as the vegetative cells (generally one of each) or whether both of these organelles were duplicated. (This literature will be reviewed later since it directly relates to the genetic significance of the statospore in the respective life cycles.) As early as 1923, Doflein observed sufficient variability in encystment among the many flagellate species he studied that he defined two types of encystment – the *Chromulina* type and the *Ochromonas* type. In *Chromulina*-type encystment, the cyst pore was formed by a rupture of the cyst membrane in the anterior portion of the cell and the resulting cysts were generally smooth, unornamented spheres. *Ochromonas*-type encystment was characterized by a cyst membrane with a preformed pore in a region of the membrane with special staining properties. These latter type cysts were often highly ornamented with elaborate pore collars or cyst body spines.

Further study of these interspecific variations was greatly enhanced by application of transmission electron microscopical (TEM) techniques. The first TEM observations of statospore contents were obtained by Wujek (1969) on the mature statospore of *Dinobryon sertularia* Ehrenberg. His micrographs depicted large storage reserves and the shrunken condition of the major organelles within the statospore which we now associate with algal and protozoan resting stages in general (Brown et al., 1968; Bowers & Korn, 1969; Neff & Neff, 1969; Bibby & Dodge, 1972; Anderson, 1975). Hibberd (1977) then published a TEM study detailing the development of the uninucleate cyst of *Ochromonas tuberculata* Hibberd. He gave an excellent account of the silicification process, development of the cyst plug, and a description of cyst maturation. His material unfortunately did not facilitate detailed study of the early stages of encystment including encystment initiation and SDV development. During the same year, two TEM papers concerning statospore formation in *Dinobryon cylindricum* Imhof (Sandgren, 1977) and *Uroglena volvox* Ehrenberg (Esser & Van Valkenburg, 1977) were presented at the Ninth International Seaweed Symposium. Most recently, I have investigated the development of the asexual, binucleate statospore of *D. cylindricum* in greater detail (Sandgren, 1980a) and have surveyed some significant stages in the encystment of *Dinobryon divergens* Ehrenberg, *U. volvox* Ehrenberg, *U. americana* Calkins, and *Mallomonas caudata* Iwanoff (Sandgren, 1980b). The results of these reports can be summarized by considering the similarities and differences that have been documented in the developmental process of encystment for the organisms studied to date. The general process of encystment as documented for *D. cylindricum* is illustrated in Fig. 1 and will be used to focus the following discussion.

The asexual, binucleate cysts of *D. cylindricum* (and most probably those of *U. volvox* as well) are initiated by an uncoupling of the nuclear and cytoplasmic events of cell division. A unique pre-encystment nuclear division occurs, resulting in the replication of the cell nucleus and chloroplast but not the other organelles (Sandgren, 1980a). This enlarged binucleate cell then undergoes encystment.

Encystment commences with both the loss of typical vegetative cell shape and some rearrangement of major organelles. The flagellar apparatus with its associated structural microtubule system moves into the interior of the cytoplasm, whereas in at least some species the dictyosome migrates into the periphery of the rounded cell (Fig. 1A). The chloroplast(s) loses its vegetative shape but remains in close association with the nucleus (nuclei) as it is still enclosed by the periplastidial endoplasmic reticulum membranes. In the binucleate cysts of *Dinobryon* and *Uroglena,* the two chloroplast/nucleus pairs assume a symmetrical position in the cell anterior to the greatly enlarged chrysolaminarin vacuole.

The SDV originates in the periphery of the anterior cytoplasm adjacent to the dictyosome. It grows to encircle the central cytoplasm with a bilayer of membrane (Fig. 1B). In all cases studied with the electron microscope, the intracystic cytoplasm consistently contains the major organelles, whereas the extracystic fraction invariably consists of a small portion of cytoplasm including some mitochondria and the contractile vacuole (Fig. 3). The TEM observations have thus confirmed the general configuration of an encysting cell as described in early light microscopical observations.

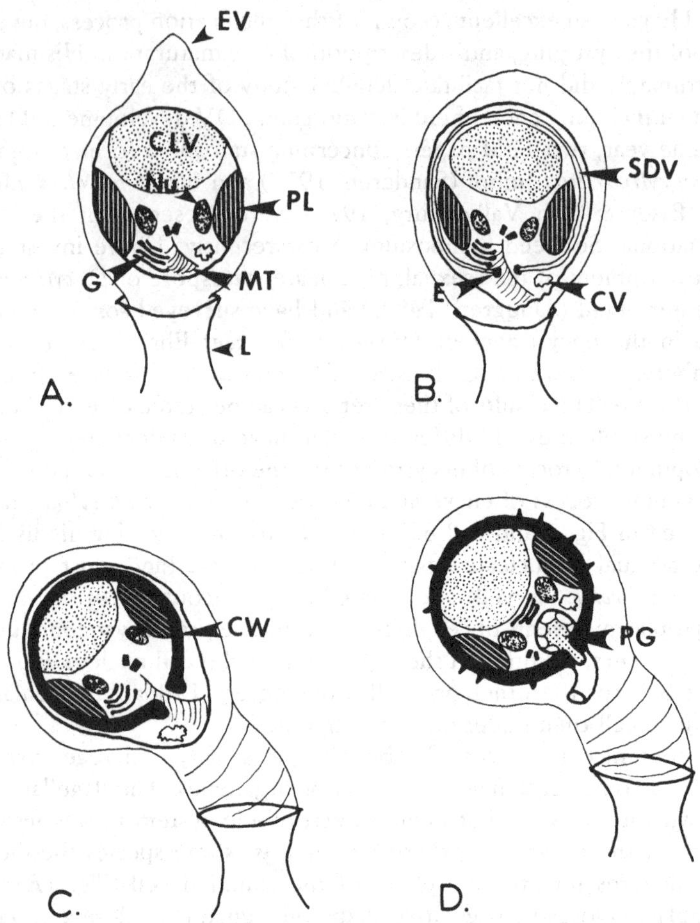

Fig. 1. Important stages in the encystment of *Dinobryon cylindricum.* (A) Early stage of statospore formation prior to SDV development. Two nuclei and two plastids are present as well as a single dictyosome, one flagellar apparatus, one contractile vacuole, and one chrysolaminarin storage vacuole. (B) An intermediate stage following the completion of SDV formation. Note position of the contractile vacuole in the extracystic cytoplasm, the electron-dense material outlining the pore in the SDV, and the microtubule system extending from the basal bodies through the pore. (C) Completion of the first stage of silica deposition. The silica cyst wall is unornamented and lacks the collar extension typical for mature cysts of this species. Extracystic cytoplasm still persists. (D) Mature statospore. A plug composed of several layers of diverse materials fills the pore. The extracystic cytoplasm has been withdrawn into the cyst, and the typical cyst ornamentation is present. Note that the cyst remains binucleate. *List of abbreviations used in EM plates:* bb, flagellar basal bodies; CLV, chrysolaminarin vacuole; CV, contratile vacuole; CW, cyst wall; E, electron-dense material; EV, encystment vesicle; G, Golgi dictyosome; Li, lipid bodies; MT, microtubules; Nu, nucleus; P, cyst pore; Pg, layers of cyst plug; Pl, plastid; SDV, silica deposition vesicle.

Early in cyst development, a pore forms in the SDV and this will eventually become the site of the pore in the mature cyst. Pore formation can apparently occur by two quite distinct mechanisms in chrysophycean flagellates. In *Dinobryon* and *Uroglena,* the SDV contains a preformed pore and so is analogous to Doflein's *Ochromonas*-type encystment pattern. The pore is outlined by a ring of electron-dense material that has a fibrous or striated substructure (Figs. 2, 3). Preliminary evidence (Hoffman, unpublished) suggests that this type of pore formation occurs in *Hydrurus foetidus* (Villars) Trevison as well (Fig. 5). Pore formation in *O. tuberculata,* on the other hand, is quite different; the pore forms by the rupturing of SDV membranes as silicification is occurring, and there is no discrete electron-dense material associated with the pore margins (Fig. 4). This mode of pore formation is thus analogous to Doflein's *Chromulina*-type pattern. Several structural similarities suggest that pore formation in *M. caudata* also follows the *Chromulina*-type pattern (Sandgren, 1980b).

All species for which early stages of encystment have been examined exhibit a system of microtubules extending from the flagellar basal bodies through the SDV pore and into the extracystic cytoplasm. These microtubules are a remnant of the structural microtubule system found in vegetative cells. The complexity of the system is dependent on the morphology of the vegetative cells themselves and so varies between species. The structure of the microtubule array of *D. cylindricum, U. volvox, O. tuberculata,* and *H. foetidus* as observed during encystment can be compared in Figs. 2, 3, 4, and 5, respectively. It appears that these microtubules play a structural role in the extracystic cytoplasm and may be involved in the retraction of this material into the developing cyst for some species (Sandgren, 1980a).

All silicification occurs within the silica deposition vesicle, and the process proceeds as two phases. The first phase results in the construction of an unornamented silica sphere, the cyst body (Fig. 1C). For species such as *U. americana* that characteristically produce morphologically unspecialized cysts, this is apparently the only phase of silicification to occur. For species such as *M. caudata, U. volvox, O. tuberculata,* and *D. cylindricum* that produce ornamented cysts, a more or less distinct second phase of silicification occurs. This second phase results in the production of structural complexities, such as spines or elaborate collars surrounding the cyst pore. New and specialized growth of the SDV membranes to serve as a template for precipitating silica must precede the second phase of silicification. No microtubules have been observed to be associated with the rapidly growing SDV at the tips of developing spines or collars, but the membrane is usually covered with a thin layer of fibrous electron-dense material (microfilaments?) indistinguishable from that surrounding

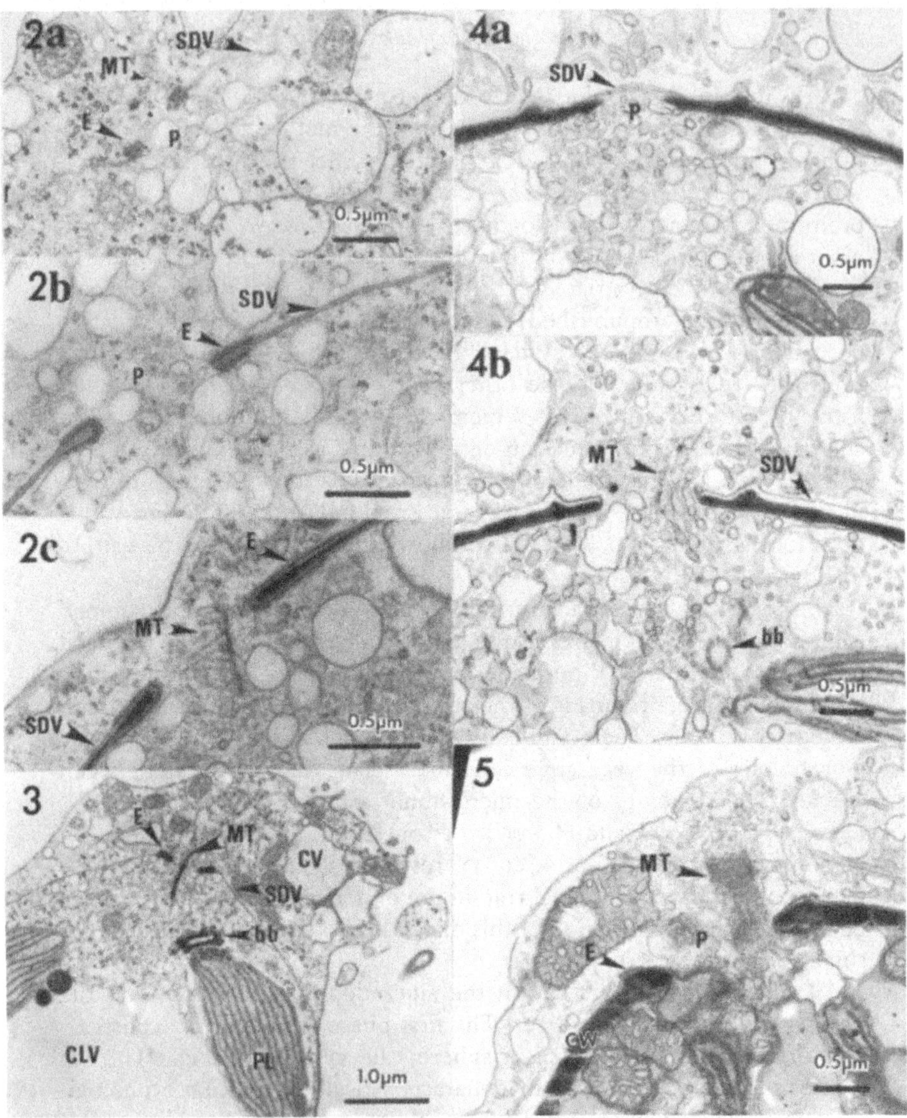

Figs. 2–5. Fig. 2. Pore formation and structure in *Dinobryon cylindricum* statospores. Pore is created as the SDV develops, and the pore margins are covered by a fibrous, electron-dense material. The cytoskeletal microtubule system extends through the pore during early stages of cyst development (arrows). (a) Pore morphology prior to silica deposition. (b) Pore morphology as silification begins. (c) Pore morphology during silica deposition. Note that the electron-dense material becomes diffuse but persists during the entire period of silification. **Fig. 3.** Pore formation and structure in *Uroglena volvox*. Note the microtubule system extending through the pore (arrows) and electron-dense material defining pore margins. **Fig. 4.** Pore formation and structure in *Ochromonas tuberculata* (micrographs from Hibberd, 1977). (a) No preformed pore is found; the pore forms by rupturing of the SDV. (b) Following pore formation, the cytoskeletal microtubules extend through the pore as in *Dinobryon*. Note that pore formation is delayed until well after the beginning of silica deposition. **Fig. 5.** Pore morphology in *Hydrurus foetidus* (Hoffman, unpublished micrograph). Note the extension of a very elaborate cytoskeletal microtubule system through the pore (arrows) and the electron-dense material associated with the pore margin.

the SDV pore in *Dinobryon* and *Uroglena* cysts (Hibberd, 1977; Sandgren, 1980a, 1980b). I have observed that this second phase of silicification may be disrupted in some cases for unknown reasons; the result being the production of both ornamented and unornamented cysts by a single population of cells (Sandgren, 1980b, 1981).

Upon completion of the cyst body wall and concurrent with or preceding the development of cyst ornamentation, some species possess the ability to retract the extracystic cytoplasm into the cyst body through the pore. The events of this process have been documented in *D. cylindricum*, and it is thought to occur similarily in *D. divergens, U. americana*, and *U. volvox. O. tuberculata* and *M. caudata*, on the other hand, do not conserve the extracystic cytoplasm. In these latter species, cytoplasm external to the cyst wall is eventually sloughed off and lost to the cell.

Formation of the plug within the cyst pore occurs during or just after the second phase of silicification (Figs. 1C,D). The structure of the plug is frequently complex and is the most variable aspect of the encystment process (aside from the often species-specific ornamentation patterns of the outer cyst surface). The plug is composed of one to three morphologically distinct layers formed from material deposited from Golgi vesicles (Fig. 6). Structural complexity of cyst plugs for all species examined with the TEM is reviewed in Sandgren (1980b).

Plug formation effectively seals off the encysted cell from the external environment. The cytoplasm then undergoes a maturation process within the cyst that involves the following events: the organelles become electron dense and shrink in volume (indeed, the whole cytoplasmic mass may shrink away from the inner wall of the cyst), there is an increased accumulation of lipid droplets, and there are striking structural changes in the contents of the chrysolaminarin vacuole (Fig. 6).

The general consistency of the encystment process throughout all chrysophycean species examined suggests that this process is an evolutionarily conservative one. If this is true, then variations between species in the process of cyst development can be used to ascertain phylogenetic relationships within the Chrysophyceae just as aspects of cell division have been used by Pickett-Heaps (1975) and Stewart and Mattox (1975) to reassess the phylogeny of green algae. I have suggested (Sandgren, 1980b) that at least two lines of evolution may be found among chrysomonad flagellates based upon variations in the encystment process. The important elements of Doflein's *Ochromonas*-type encystment pattern are found among members of the Dinobryaceae (*Dinobryon*) and certain advanced members of the Ochromonadaceae (*Uroglena*). Critical aspects of his *Chromulina*-type encystment pattern are found in some ochromonad unicells (*O. tuberculata*) and perhaps in the Synuraceae as exemplified by *M. caudata*. This phylogenetic proposal is a preliminary one, but it will hopefully stimulate additional investigations of statospore de-

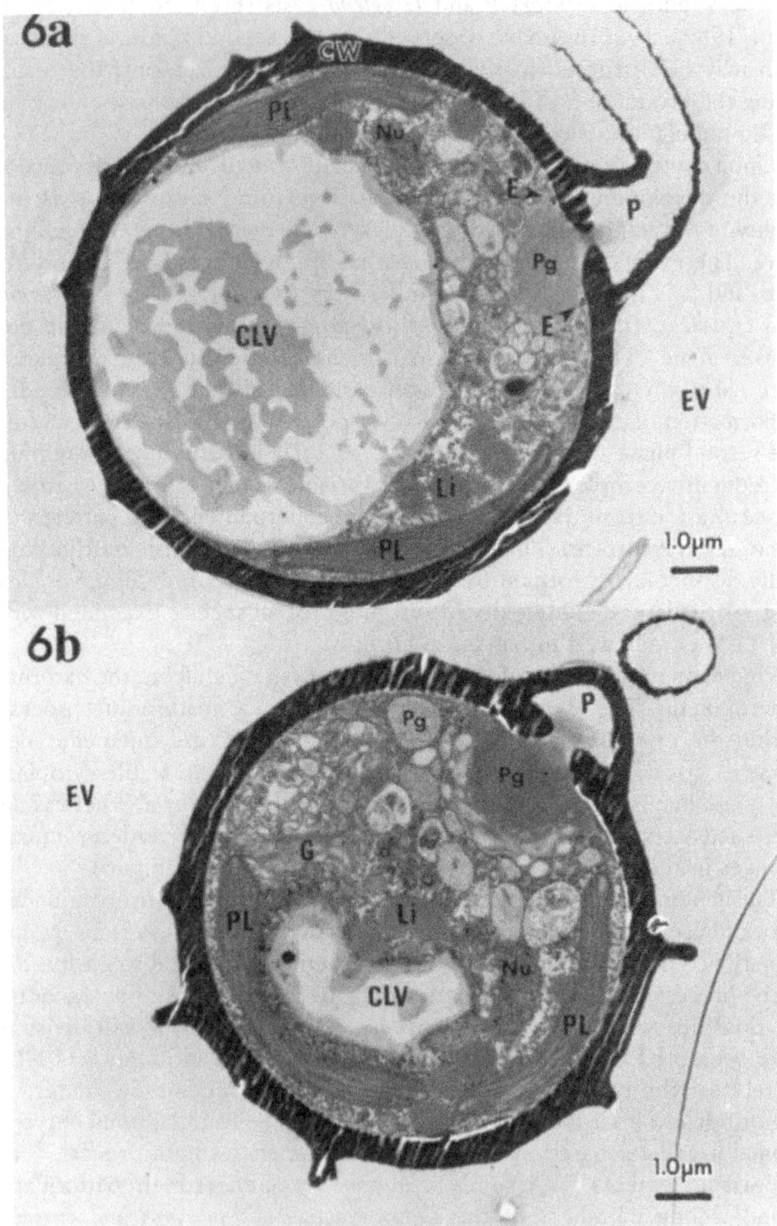

Fig. 6. Mature cyst of *Dinobryon cylindricum*. (a) Cross section showing the typical arrangement of organelles. The cytoplasm and major organelles are shrunken and electron dense in appearance. There are numerous lipid bodies in the cyst cytoplasm, and the chrysolaminarin has undergone dramatic structural alterations as compared to the vegetative cell. (b) Cross section 90° rotated from (a).

velopment in a variety of chrysophycean species. Such research may help to clarify the phylogenetic relationships within this division of the algae.

2.2 Cyst surface morphology

The statospores of many chrysophycean species are elaborately ornamented with species-specific patterns including highly characteristic cyst collars as well as a great variety of projections from the cyst surface. There are many light microscopical descriptions of statospore ornamentation scattered in the literature, and for some flagellate genera (e.g., *Uroglena* and some species of *Ochromonas* and *Chromulina*) a knowledge of statospore morphology is necessary to identify species.

The details of statospore ornamentation can be very effectively determined with the scanning electron microscope (SEM), but there have been few published SEM micrographs of cyst morphology (Cronberg, 1973, 1980; Gritten, 1977; Hibberd, 1977). Some of the variety in ornamentation can be observed from the micrographs presented here (Figs. 7–12). The cyst surface may consist of a solid lamina of silica representing the cyst body wall (Fig. 7), or this basal sphere may be completely covered by a tightly packed "forest" of reticulate elements as revealed by thin sections of *M. caudata* cysts (Sandgren, 1980b). Solitary projections from the cyst body may consist of stubby spines (Figs. 8, 9), greatly elongate and sometimes bifurcate processes (Fig. 10), or ridges that may be variously described as rugulose or crenulate (Fig. 11). The very striking cysts of *Chrysidiastrum catenatum* Lauterborn (Fig. 12) are ornamented with arcuate recessed pockets and several much larger posterior cavities.

Because statospores are siliceous, they are as readily preserved as fossils in lake sediments as are diatom frustules. Several attempts to use chrysophycean cysts or the vegetative cell scales of members of the Synuraceae as paleolimnological indicators have been made (Tippet, 1964; Elner & Happey-Wood, 1978; Munch, 1980; Smol, 1980), but the number is very small relative to the great number of paleodiatom publications. I believe that the use of these highly characteristic chrysophycean microfossils has been neglected by paleolimnologists chiefly because the cysts preserved in sediments cannot be readily correlated with the vegetative cells that form them. There is a lack of detailed descriptions for statospores that would allow them to be definitely assigned to individual chrysophycean species. Instead, an artificial taxonomy of form genera has been created for statosporelike microfossils (Conrad, 1939; Deflandre, 1952; Nygaard, 1956). Until the link between these fossilized cysts and the free-living chrysophycean species that produce them can be made by phycologists, it is unlikely that statospores will assume an

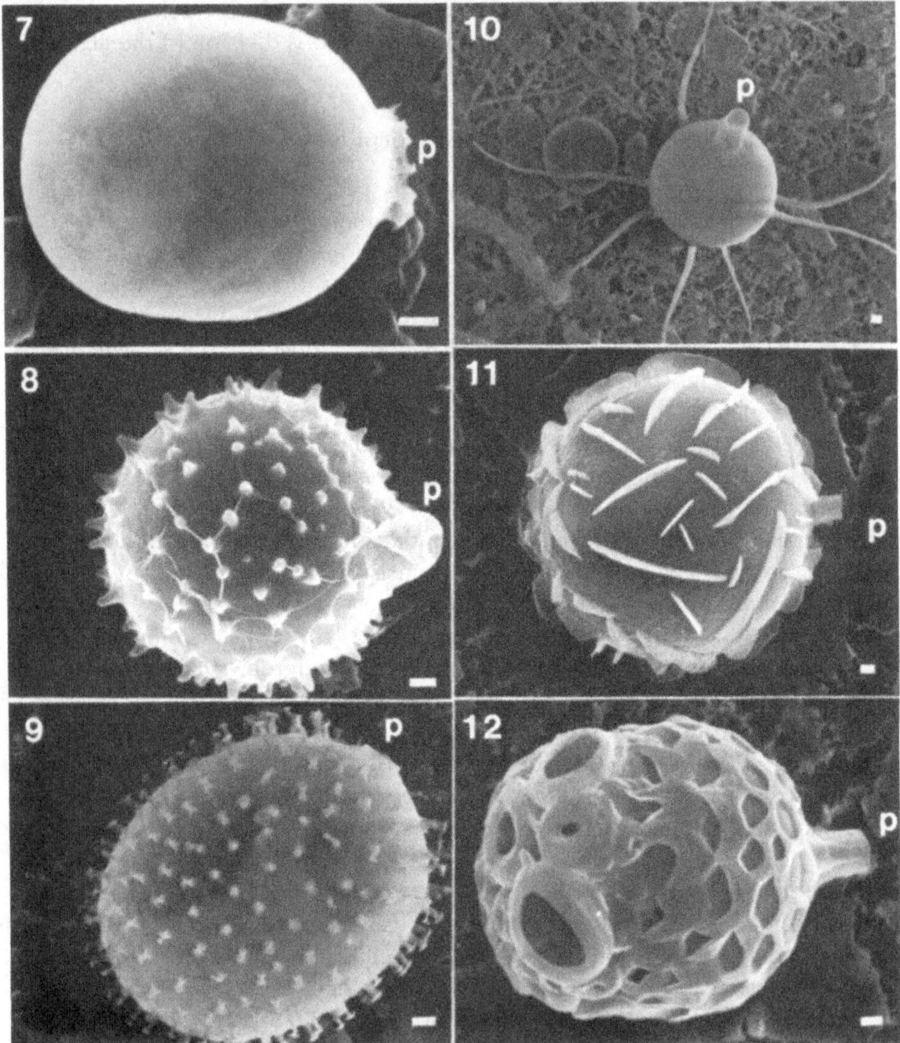

Figs. 7–12. SEM micrographs of mature chrysophycean statospores. All scale bars = 1.0 µm. **Fig. 7.** *Mallomonas akrokomonas*. **Fig. 8.** *Dinobryon cylindricum*. **Fig. 9.** *Mallomonas torquata* Asmund and Cronberg 1979 (see Cronberg, 1980). **Fig. 10.** *Mallomonas* cf. *fastigata*. **Fig. 11.** Unidentified *Ochromonas* sp. **Fig. 12.** *Chrysidiastrum catenatum*.

important role as paleoecological indicators. I believe they have great potential in this regard.

2.3 Significance of statospores in the life history of chrysomonads

The principal function of the statospore in the chrysophycean life history is to ensure the survival of the species between periods of vegetative cell growth (Gayral et al., 1972). The statospore's resistant siliceous walls, the large reserves of carbohydrates and lipids, and the shrunken, inactive appearance of the cellular organelles as described by electron microscopical observations clearly support this conclusion. However, the genetic significance of the statospore is uncertain for most species, and these cysts may well prove to be multifunctional for many.

Several reports document the simultaneous development of morphologically identical or very similar cysts by both sexual and asexual processes in natural populations of some planktonic chrysophycean species (Skuja, 1950; Bourrelly, 1957; Fott, 1964; Kristiansen, 1965). In addition, there are numerous, seemingly contradictory descriptions of either uni- or binucleate cysts for a number of *Dinobryon* species (uninucleate: *D. crenulatum*, Asmund, 1955; *D. divergens*, Entz, 1930; Sheath et al., 1975; *D. sertularia*, Meyer, 1965; binucleate: *D. cylindricum*, Krieger, 1930; Sandgren, 1978, 1980a; *D. divergens*, Krieger, 1930; Geitler, 1935; Sandgren, 1980b). Indeed, lengthy lists of both uni- and binucleate statospores produced by various chrysophycean species have been assembled (Mack, 1951; Sandgren, 1980b). Because many of the reports of binucleate cysts concern loricate chrysomonads of the family Dinobryaceae, and because this family is also distinctive in that almost all unequivocal reports of gametic sexual reproduction concern its members (as reviewed in Bourrelly, 1957, 1963; Fott, 1959, 1964; Kristiansen, 1961, 1963), it is tempting to infer that binucleate cysts are always indicative of gametic sexuality. However, observations of binucleate *D. cylindricum* cysts produced as the result of a preencystment nuclear division of the uninucleate cyst mother cell (Sandgren, 1980a) preclude such a sweeping conclusion as do recent reports of apparent autogamy in an *Ochromonas* sp. (Gayral et al., 1972).

One point of consistency exists concerning the genetic signficance of the statospore. In all reports of gametic sexual fusion in the Chrysophyceae, the resulting zygospore has been a statospore having the morphology typical for the species involved (Fott, 1959). Fusion of vegetative cells (hologamy) has been observed in natural populations of *Chrysolykos planctonicus* Mack, *Dinobryon borgei* Lemmermann, *Kephyrion mastigophorum* Schmid, *K. rubri-claustri* Conrad, *K. translucens* Fott, *Kephyriopsis conica* Schiller, *K. cylindrica* (Lackey) Fott, *K. cincta* Schiller, *K. entzii* (Conrad) Kristiansen, *Kephyriopsis* sp. (Mack, 1954), *Mallomonas caudata* Iwanoff, *Stenocalyx inconstans* Schmid, *S. klarnetii* (Bourrelly) Fott, *S. monilifera* Schmid, *S. spirale* (Lackey) Fott, *S. tubiforme* (Fott) Fott, and *Synura petersenii* Korschikov. In all cases, the fusing vegetative cells are described as being morphologically isogamous. The inference is that chrysophycean cells are vegetatively haploid and that the statospore serves as a resting zygospore.

There is no doubt, therefore, that the statospore can be a sexually induced resting stage, but the many examples of binucleate cysts appar-

Table 1. *Encystment resulting from binary clone crosses:* Dinobryon cylindricum *Imhof*

Clone	1	3	5	7	11	13
A. Exponential growth phase (Day 7)						
1-	0.047	0.011	0.063	10.0	0.020	5.3
3-		0.005	0.001	0	0.001	0
5-			0	29.9	0.001	15.1
7-				0.003	0	0
11-					0	0
13-						0
B. Stationary growth phase (Day 17)						
1-	0.027	0.117	0.057	16.36	1.09	10.49
3-		0.002	0.360	0.001	0.004	0.003
5-			0.007	22.61	1.36	12.74
7-				0.002	0.003	0.011
11-					0.001	0.001
13-						0.001

Note: Results of binary clone crossing experiments involving all combinations of six *D. cylindricum* clones under controlled laboratory conditions. Results are expressed as an encystment frequency (= % cysts/cell) and are tabulated during both the exponential (Day 7; Part A) and stationary (Day 17; Part B) growth periods for the same three replicates of each cross. Mean encystment frequencies are listed. Encystment frequencies of the six clones when grown alone are found in the first diagonal of the data matrices. The source of the clones, some physiological data, and the experimental conditions can be found in Sandgren (1981).

ently produced without prior gamete fusion reported above and the many reports of uninucleate cysts developing from single vegetative cells (Doflein, 1923; Geitler, 1927; Korschikov, 1929; Krieger, 1930; Skuja, 1950; Mack, 1951, 1954; Bourrelly, 1957; Fott, 1964, Kristiansen, 1965; Hibberd, 1977; Sandgren, 1980b) strongly suggest that the statospore may more often be an asexual resting stage. I have recently completed a series of culture experiments designed to elucidate the genetic significance of statospores produced by the widely distributed planktonic chrysomonad, *D. cylindricum* (Sandgren, 1981).

The results of binary clone crosses of *D. cylindricum* (Table 1) suggest that this species is capable of producing several genetically distinct types of statospores. A clone such as clone 1 produces cysts typical of *D. cylindricum* when it is grown alone under standard conditions. Under these conditions of optimal light, temperature, and nutrients, such a clone produces cysts at a relatively low cyst/cell frequency (f_c = 0.05%). Feulgen staining reveals these to be an approximately equal mixture of uninucleate and binucleate cysts, which can be separated only imper-

Table 2. Dinobryon cylindricum *statospore morphology*

Source of statospores	Diameter (μm)	No. of plastids	No. of nuclei (Feulgen stain)
Clone 1	10–12	1	1
	14–16	2	2
Clone 2	14–16	2	2
Clone 11	14–16	2	2
Clone crosses			
3 × 5	14–16	2	2
1 × 7	14–16	2	2
1 × 13	14–16	2	2
5 × 7	14–16	2	2
5 × 13	14–16	2	2
Egg Lake plankton, 1978	14–16	2	2
Lakedale plankton, 1978	14–16	2	2

fectly on the basis of size (Table 2). Both ornamented (Fig. 8) and unornamented cysts are found in both size classes.

Some binary crosses of clones (1 × 7, 1 × 13, 5 × 7, 5 × 13) produce cysts at comparatively high frequencies (f_c = 5%–30%) during exponential growth phase of the cultures (Table 1A). Other crosses (1 × 3, 1 × 11, 3 × 5, 5 × 11) produce cysts at lower frequencies (f_c = 0.1%–1.4%), but these values are still significantly higher than the encystment frequency of any clone grown alone. In these later four crosses, cysts are only produced in numbers after the populations have reached high cell densities during stationary growth phase (Table 1B).

In all eight crosses that result in the production of relatively high numbers of cysts, the clones involved produce no or very few cysts when grown individually. Cysts produced by crossing these clones are binucleate and are morphologically indistinguishable from the large binucleate cysts produced by clone 1 individually (Table 2). Careful observations of clonal crosses involving high-frequency encystment (1 × 13, 5 × 7) have confirmed the existence of gametic fusion resulting in zygotic statospore formation in these cases (Sandgren, 1981). One gamete is a typical loricate cell; the second gamete is a nonloricate cell produced as the result of a recent cell division occurring in a loricate cell of the compatible clone.

A series of experiments has been carried out on these sexually compatible clones using U-shaped culture tubes in which the two arms of the tube can be separated by a Millipore filter of 0.45-μm pore size. These experiments have more clearly documented the sexual process by separating chemical and cell-contact events. When compatible clones are in chemical contact across the filter boundary, the male clone (clone 5) is

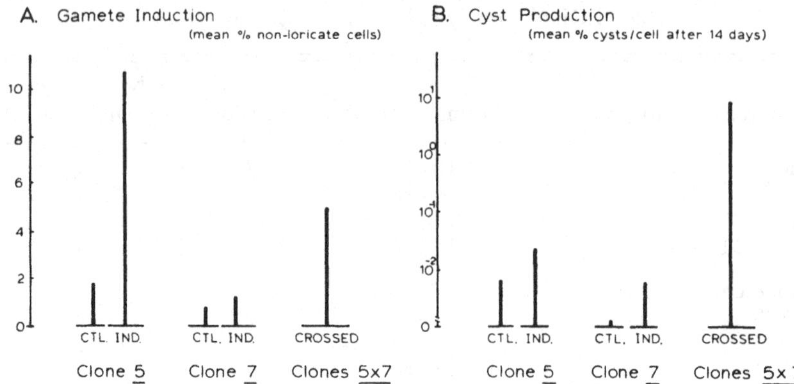

Fig. 13. An example of the experiments used to examine the chemical and physical events involved in sexual statospore formation in *Dinobryon cylindricum.* Sexually compatible clones (clones 5 and 7) were grown for 10 days in U-shaped culture vessels in which the two arms were separated by a 0.45-μm Millipore filter. Compatible clones were grown in separate U tubes (control conditions, CTL.), in opposing arms of the same U tube (inducing conditions, IND.), or mixed in equal proportions within a single arm (crossed condition, CROSSED). The results are presented both in terms of the percentage of nonloricate male gametes produced (A) and in terms of the percentage cysts produced (B). One-way analysis of variance and Student–Newman–Keul multiple-comparison tests confirm that nonloricate gamete production is significantly enhanced under induced and crossed conditions and that cyst production is enhanced only under crossed conditions (α-0.05). Note logarithmic scale for % statospore production in B.

induced to form a high percentage of nonloricate gamete cells, an event not observed in the female clone (clone 7) (Fig. 13A). However, no increase in statospore formation occurs over that of the control cultures (Fig. 13B). Statospores are only produced in abundance when compatible clones are in mixed culture within one arm of the U tube and thus are in both physical and chemical contact (Fig. 13B, CROSSED conditions).

On the basis of these experiments, sexual reproduction in *D. cylindricum* apparently involves a diffusable gamete-inducing substance produced by female clones (clones 3, 7, 11, 13), which results in the appearance of nonloricate male gametes from compatible male clones (clones 1, 5). Compatibility appears to be a relative phenomenon since the percentage of encysting cells and the size of the vegetative cell population necessary to induce encystment in compatible clones are quite variable. Both compatible and noncompatible clones can coexist in mixed natural populations (Sandgren, 1978, 1981).

Until the events of excystment are documented, the genetic significance of the binucleate cysts produced by single clones such as clone 1 remains unknown. These may be autogamic as suggested by Geitler (1935). Mack (1954), Fott (1959), and Gayral et al. (1972); the clone involved would thus be self-compatible. However, autogamy is of little value in terms of genetic recombination for organisms such as these that are vegetatively haploid, and the evidence for eventual fusion of the two nuclei in "autogamic" cysts is meager (Prowazek, 1903, Dangeard, 1910; Mack, 1951). Because the binucleate condition is not required to initiate the encystment process, autogamy does not appear to have any selective value for chrysophycean species on the basis of our present understanding.

Culture studies have therefore confirmed all the surprising diverse functions proposed for the statospore within the chrysophycean life history. Using *D. cylindricum* as an example, encystment can be of several types. It can involve individual cells of a single clone producing either uni- or binucleate asexual cysts of similar morphology but slightly different diameter, or it can involve two clones representing compatible mating types producing sexual cysts at varying encystment frequencies depending on the degree of compatibility of the clones involved. I find it rather amazing that all these potentially genetically distinct types of statospores can be produced by clones of a single planktonic species. The diversity of triggering mechanisms that lead to morphologically similar statospores emphasizes the complexity and evolutionarily conservative nature of the encystment process. This diversity also emphasizes the extreme importance of these resistant resting cysts to the continued survival of natural chrysophycean populations.

2.4 Factors influencing encystment

An understanding of the factors influencing the rate of encystment are of basic importance in understanding the reproductive strategies of chrysophycean flagellates. Because cysts may be essential for the persistence of the species, factors influencing the rate of encystment and thus the total size of the resulting cyst population will certainly influence the survival of the species in a given environment. It has been established by the culture experiments outlined here that the nature of the clones making up any natural population will certainly influence the reproductive success of that population. Self-compatible or asexually encysting clones could exist as monotypic populations in nature, whereas clones requiring a compatible mating type in order to form statospores could not unless the environment were very stable. These same culture experiments further suggest that perhaps all clones produce some cysts, but for many the encystment frequency is very low (1 cyst/10^5 cells). The potential significance of a range of encystment frequencies in terms of the resulting size of the cyst population will be further considered for natural chrysophycean populations in the next section.

There are no published accounts of the effects of environmental stresses on the rate of statospore production although an effect of temperature has been proposed (Fott, 1964). Examination of the effect of

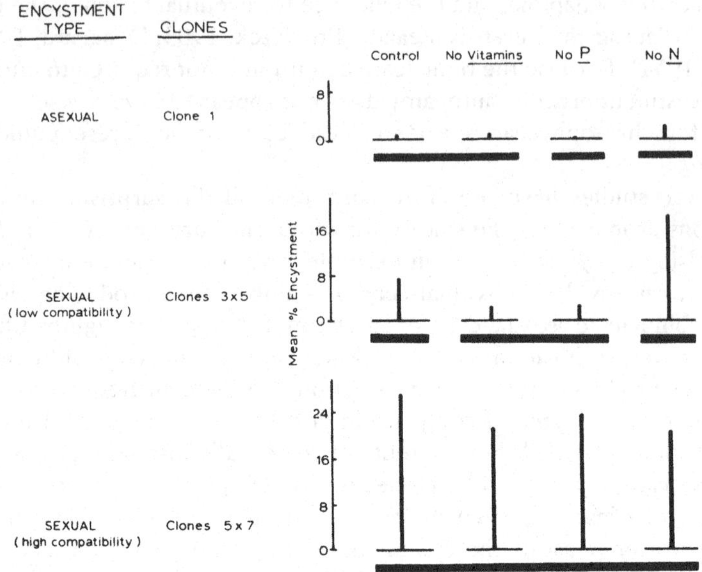

Fig. 14. Effect of nutrient stress on the rate of statospore production for the three types of statospores documented for *Dinobryon cylindricum*. Results are expressed as an encystment frequency (% cysts/cell). Cultures were grown under standard conditions of 100 μE/m²/sec (PAR) cool white fluorescent light, 16-hr light: 8-hr dark cycle, and 16–19°C. Horizontal bars beneath histograms denote the results of one-way analysis of variance and Student–Newman–Keuls multiple-comparison tests on replicated counts from three replicate cultures in each manipulation. The experiment was terminated early in the stationary growth phase (Day 13) so that no confounding of results by the rapid decline in cell density following severe nutrient limitation could occur. No significant differences in the final yield of vegetative cells was detected for any manipulation of the three types of encystment (one-way analysis of variance) except for nitrogen stress of clone 1, where vegetative growth appeared enhanced.

several nutrient deficiencies on the production of the three types of *D. cylindricum* statospores (Sandgren, 1981) shows that asexual encystment of clone 1, which includes the production of both uninucleate and binucleate cysts, is greatly enhanced by nitrogen stress and is inhibited by phosphorus stress (Fig. 14). Low-compatibility sexual encystment (clones 3 × 5) is also greatly enhanced by nitrogen stress but is inhibited by the deficiency of vitamins (B_1, B_6, B_{12}) as well as of phosphorus. In contrast, high-compatibility sexual reproduction (clones 5 × 7) is not affected by any of the nutrient deficiencies created. The dynamics of cyst production (Sandgren, 1981) suggest that encystment is continuous at the expressed frequency during the entire exponential and linear (or early stationary) growth periods for clone 1 and clones 3 × 5. Some high-compatibility sexual crosses (1 × 7, 1 × 13) produce cysts primarily during the exponential growth period (*intrinsic* encystment), whereas in other crosses (5 × 7, 5 × 13) encystment is delayed until the linear growth period (*extrinsic* encystment). The basis for this divergence with respect to encystment dynamics has not yet been determined, but it certainly is not due to nutrient stress experienced during the stationary growth phase of culture.

It is concluded that asexual (and/or autogamic) encystment of *D. cylindricum* occurs at low frequency (0.01–1.0 [5.0]%) under all of the environmental conditions examined, whereas high-compatibility sexual reproduction results in a consistently high encystment frequency (>10%). Low-compatibility sexual encystment proceeds at an intermediate frequency (2%–7%) but can be greatly enhanced by nitrogen stress. Encystment of natural chrysophycean populations might be expected to vary widely with respect to encystment frequency, the dynamics of encystment, and the effects of nutrient stress if the results of these *D. cylindricum* culture experiments are representative of encystment characteristics within the class Chrysophyceae as a whole.

2.5 Encystment of natural chrysophycean populations

General limnological studies that include documentation of population dynamics for chrysophycean species seem inevitably to omit any quantitative estimates of statospore formation (Ruttner, 1930; Vetter, 1937; Hutchinson, 1967). Much of the available information concerning the encystment of natural chrysophycean populations thus consists of observations from either a single sampling date or from several dates that are widely spaced in time. Very few field studies have consistently monitored the vegetative growth and cyst production of golden algal species with sufficient frequency to establish dynamic patterns of encystment in nature (Cronberg, 1973; Sheath et al., 1975; Sandgren, 1978). On the basis of these few studies, however, it is apparent that the encystment dynamics of populations in nature varies widely between species and among populations of a single species.

Encystment of 34 planktonic chrysophycean populations occurring in Egg and Sportsmans Lakes, San Juan Island, Washington, was carefully followed using a 2 to 3-day sampling regime in the spring of 1976 and 1977 (Sandgren, 1978, and unpublished data). In many instances, the observed period of cyst production was very brief or of such low encystment frequency that it could easily be overlooked in standard phytoplankton sampling programs with a periodicity of 1 week or more. Both intrinsic and extrinsic patterns of encystment as defined with respect to *Dinobryon* culture experiments were observed, and the range of encystment frequency was broad. The two most common encystment patterns are those exemplified by *D. cylindricum* and *Synura spinosa* Korschikov (Fig. 15). Species such as *D. cylindricum* for which vegetative growth appears restricted to short periods of time (1–3 weeks) often produce

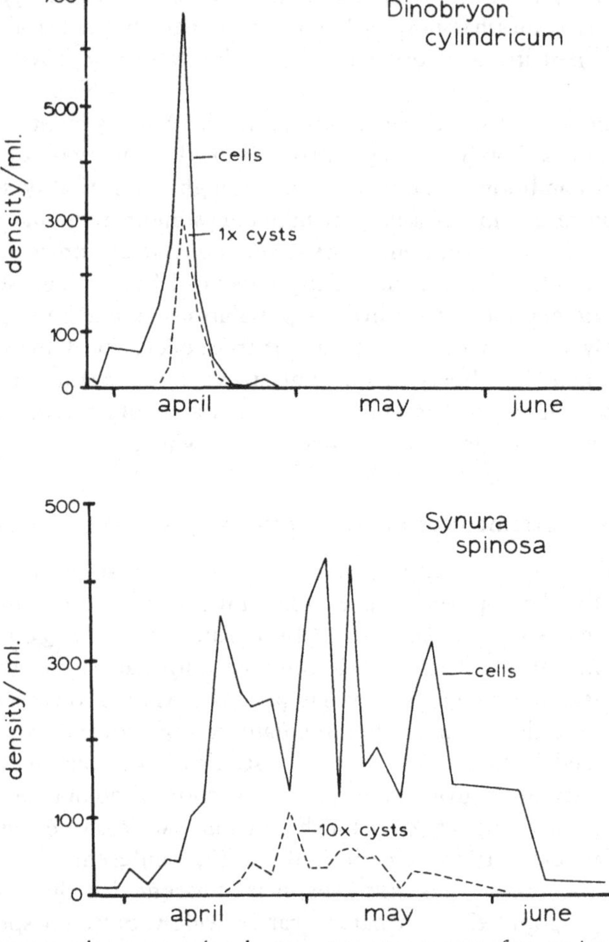

Fig. 15. Two examples representing the most common patterns of vegetative population growth and statospore (cyst) formation observed among 34 natural chrysophycean populations studied in Egg and Sportsmans Lakes, 1976 and 1977. Note that cyst density of *Synura spinosa* is exaggerated by ten times its actual value.

cysts at high mean frequencies (20%–30%), exhibiting either extrinsic or intrinsic encystment dynamics. Species such as *S. spinosa* that maintain standing populations for several months most commonly produce cysts at low encystment frequencies (<5%), but the episode of encystment follows the intrinsic pattern and may continue for many weeks. In general, the dynamics of the spring chrysophycean species can be summarized in saying that the assemblage was dominated by sustained populations of several *Mallomonas* sp. (*M. acaroides* Perty emendavit Iwanoff, *M. akrokomonas* Ruttner, *M. crassisquama* (Asmund) Fott) and *Synura spinosa*, all exhibiting low-frequency intrinsic encystment, while at the

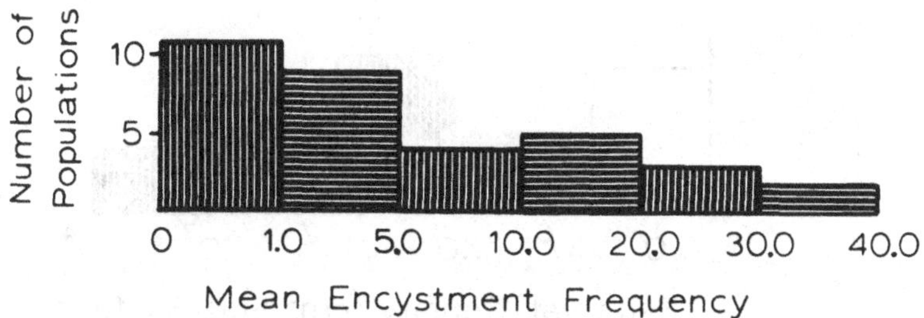

Fig. 16. Frequency diagram showing the distribution of values of *encystment frequency* among 25 populations of encysting chrysophyte species in Egg and Sportsman Lakes, 1976 and 1977. Values plotted are means calculated for the entire encystment period.

same time a regular succession of more environmentally restricted species (*D. cylindricum, D. divergens, U. volvox, Pseudopedinella erkensis* Skuja) was occurring, and many of these exhibited high-frequency encystment.

The mean encystment frequency for these 34 populations varied from 0 to 36% (Fig. 16). Nine populations were never observed to form statospores, but all of the 13 species studied did form statospores during 1 of the 2 years of observation. If encystment frequency of natural populations is to some extent indicative of the genetic significance of the resting cysts as appears to be true for *D. cylindricum* in culture, then both sexual and asexual statospores are being produced by these vernal species. Some species produce both types; many appear to produce only asexual (low frequency) statospores.

Mean encystment frequency together with the observed length of the encystment period in the plankton can be used to estimate the total size of the cyst population eventually deposited on the lake sediments by each species, irrespective of the pattern of encystment (Fig. 17). For the 25 populations in these two lakes that did form cysts, the total cyst population produced ranged from 0 to 6,440 cysts/cm^2 sediment surface with a mean density of 501 cysts/cm^2 sediment surface and a median density of 63 cysts/cm^2 sediment surface. In Egg Lake, which has a surface area of 3.5 hectares and a regular basin morphology (maximum depth, 5 m), the estimated total number of cysts deposited in the lake for the minimum and maximum density estimates would be 2.65×10^8 and 1.7×10^{12}, respectively.

These minimum and maximum estimates for the total size of the refuge populations being deposited on the Egg Lake sediments are both enormous. Unless there is very severe mortality among cysts while they reside on the sediment surface (a topic about which nothing is known), it is difficult to believe that the difference between these two densities is significant in terms of the potential to eventually repopulate the planktonic habitat. If this is true, it implies that any of the observed patterns of

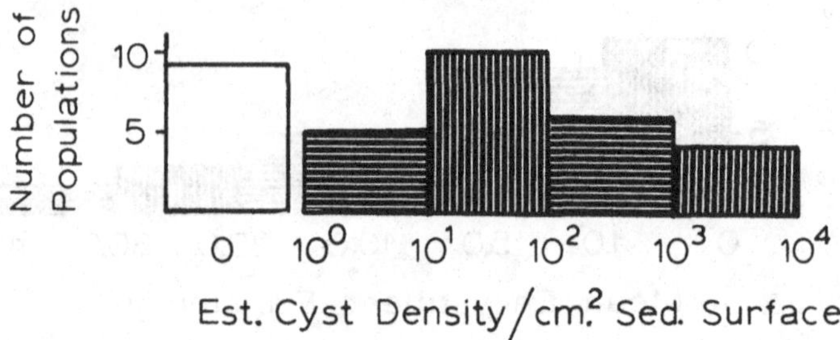

Fig. 17. Frequency diagram showing the distribution of values of *total cyst population size* produced by 34 populations of chrysophycean species in Egg and Sportsmans Lakes, 1976 and 1977. Estimates were derived by integrating mean epilimnetic cyst density for each species on each sampling date over the entire period of observed encystment. All cysts produced on a given sampling date were assumed to have sunk out of the epilimnion (2.5 m depth) by the next sampling date, 2–3 days later. Three estimates of cyst density were available for each sampling date.

encystment are adequate to ensure the establishment of a very large refuge population. Indeed, a nonparametric statistical test (rank correlation analysis) suggests that there is no correlation between mean encystment frequency and the estimated total density of the resulting cyst population (r_s = 0.17, not significant at the 95% confidence level). It would appear that if sexual reproduction is correlated with high-frequency encystment as the *Dinobryon* culture experiments suggest, then the advantage of sex for these species must be based on the enhanced rate of genetic recombination, not on the increased rate of statospore production.

Simultaneous germination of these large cyst populations could result in an explosive reappearance of chrysophycean series in the plankton, and this is often observed in nature. Studies of the benthic survival of chrysophycean cysts and the conditions that promote germination are essential for understanding the population dynamics of these species. One hopes these will become an active area of phytoplankton research. It is certainly conceivable that much of the population dynamics of phytoplankton that we interpret as seasonal succession patterns are, in fact, a reflection of species-specific differences in the germination strategies of benthic resting cell populations.

2.6 Conclusion

The last 5 years have brought great advances to the understanding of the development and function of the chrysophycean statospore. However, the picture is still incomplete and biased by a reliance on information from only a few species.

Many avenues of research in chrysophycean life history strategies are now open. There is a need for additional TEM studies of statospore formation to corroborate or disprove the proposed phylogenetic significance of variations in the encystment process. Scale-covered chrysomonads (Synuraceae), naked chrysomonads (Chromulinaceae, Ochromonadaceae), and members of orders other than the Chrysomonadales are of particular importance in this regard. Descriptive SEM studies of statospore morphology, which can be used to correlate fossilized cysts from lake sediments to the vegetative cells that have formed them, are of great potential importance to paleolimnologists and paleoecologists. The existence of distinct mating types and mating groups in *D. cylindricum* suggests that the complexity of the sexual mating systems in the Chrysophyceae may be as great as that documented for volvocalean green algae and ciliate protozoans. A great deal of research on chrysophycean mating systems including *Dinobryon* as well as other genera remains to be undertaken before the reproduction of natural chrysophyte populations can be understood and predicted. The characteristics of statospore dormancy and eventual excystment are completely unknown. With the ability we now possess to culture chrysophycean species and induce the formation of statospores under defined conditions, these characteristics should be documented. Finally, the life history strategies of natural chrysophycean populations need to be further investigated using techniques of quantitative limnology and population ecology. An understanding of the dynamics of vegetative growth and statospore production in combination with the environmental and genetic constraints on these processes holds great promise of being able to predictively model these natural populations.

References

Anderson, O. R. (1975). Ultrastructure and cytochemistry of resting cell formation in *Amphora coffaeformis*. *Journal of Phycology,* 11, 272–281.

Asmund, B. (1955). Occurrence of *Dinobryon crenulatum* Wm. et G. S. West in some Danish ponds and remarks on its morphology, cyst formation and ecology. *Hydrobiology,* 7, 75–87.

Bibby, B. T., & Dodge, J. T. (1972). Encystment of a freshwater dinoflagellate: a light and electron-microscope study. *British Phycological Journal,* 7, 85–100.

Bold, H. C., & Wynne, M. J. (1978). *Introduction to the Algae: Structure and Reproduction.* Englewood Cliffs, N.J.: Prentice–Hall.

Bourrelly, P. (1957). Recherches sur les Chrysophycées: morphologie, phylogénie, systématique.*Revue Algologique Memoires Hors-série,* 1, 1–412.

– (1963). Loricae and cysts of the Chrysophyceae. *Annals of the New York Academy of Science,* **108**, 421–429.

– (1968). *Les Algues d'eau douce. Initiation à la systématique. II. Les algue jaunes et brunes, Chrysophycées, Phaeophycées, Xanthophycées et Diatomées.* N. Boubée, Paris.

Bowers, B., & Korn, E. D. (1969). The fine structure of *Acanthamoeba Castellanii* (Neff strain). II. Encystment. *Journal of Cell Biology,* 41, 786–805.

Brown, R. M., Johnson, C., & Bold, H. C. (1968). Electron and phase-contrast microscopy of sexual reproduction in *Chlamydomonas moewusii*. *Journal of Phycology*, 4, 100–120.
Cienkowski, L. (1870). Über Palmellaceen und einige Flagellaten. *Archiv für Mikroskopische Anatomie*, 6, 421–438.
Conrad, W. (1939). Notes protistologiques. VI. Kystes de Chrysomonadines ou Chrysostomatacees? *Bulletin du Musée Royal d'Histoire Naturelle de Belgique*, 14, 1–6.
Cronberg, G. (1973). Development of cysts in *Mallomonas eoa* examined by scanning electron microscopy. *Hydrobiologia*, 43, 29–38.
– (1980). Cyst development in different species of *Mallomonas* (Chrysophyceae) studied by scanning electron microscopy. *Archiv Hydrobiologie Supplement*, 56, 421–434.
Dangeard, P. A. (1910). Études sur le développmenr et la structure des organisms inferieurs. 3. Les flagellés. *Le Botaniste*, 11, 113–180.
Deflandre, G. (1952). Chrysomadines fossiles. In *Traité de Zoologie*, ed. P.-P. Grassé, Vol. 1, pp. 560–570. Paris: Masson.
Doflein, F. (1923). Untersuchungen über Chrysomonadinen. 3. Arten von *Chromulina* und *Ochromonas* aus dem badischen Schwarzwald und ihre Cystenbildung. *Archiv für Protistenkunde*, 46, 267–344.
Drum, R. W., & Pankratz, H. S. (1964). The post-mitotic fine structure of *Gomphonema parvulum* Kütz. *Journal of Ultrastructure Research*, 10, 217–223.
Elner, J. K., & Happey-Wood, C. M. (1978). Diatom and chrysophycean cyst profiles in sediment cores from two linked but contrasting Welsh Lakes. *British Phycological Journal*, 13, 341–360.
Entz, G. (1930). Phaenologische Aufzeichnungen und einige morphologische Beobachtungen an Chrysomonaden. *Folia Cryptogamica*, 7, 669–742.
Esser, S. C., & Van Valkenburg, S. D. (1977). The fine structure of vegetative and statospore forming cells of *Uroglena volvox* Ehrenberg. *Journal of Phycology*, 13 (Suppl.), 20.
Fott, B. (1959). Zur Frage der Sexualität bei den Chrysomonaden. *Nova Hedwigia*, 1, 115–130.
– (1964). Hologamic and agamic cyst formation in loricate chrysomonads. *Phykos*, 3, 15–18.
Fritsch, F. E. (1935). *Structure and Reproduction of the Algae*, Vols. I, II. Cambridge: Cambridge University Press.
Gayral, P., Haas, C., & Lepailleur, H. (1972). Alternance morpholigique de générations et alternance de phases chez les Chrysophycees. *Memoires de la Societé Botanique de France*, 1972, 215–230.
Geitler, L. (1927). Die Schwarmer und Kieselzysten von *Phaeodermatium rivulare*. *Archiv für Protistenkunde*, 58, 272–280.
– (1935). Über zweikernige Cysten von *Dinobryon divergens*. *Österreichische Botanische Zeitschrift*, 84, 282–286.
Gritten, M. M. (1977). On the fine structure of chrysophycean cysts. *Hydrobiologia*, 53, 239–252.
Haye, A. (1930). Untersuchungen über *Dinobryon divergens*. *Archiv für Protistenkunde*, 72, 293–302.
Hibberd, D. J. (1977). Ultrastructure of cyst formation in *Ochromonas tuberculata* (Chrysophyceae). *Journal of Phycology*, 13, 309–320.
Hollande, A. (1952). Classe des Chrysomonadines. In *Traité de Zoologie*, ed. P.-P. Grassé, Vol. 1(1), pp. 471–560. Paris: Masson.
Hutchinson, G. E. (1967). *A Treatise on Limnology*, Vol. II. Sommerset: Wiley.

Korschikov, A. A. (1929). Studies on the Chrysomonads, 1. *Archiv für Protistenkunde,* 67, 253–290.

Krieger, W. (1930). Untersuchungen über Plankton-Chrysomonaden. Die Gattungen *Mallomonas* und *Dinobryon* in monographischer Bearbeitung. *Botanisches Archiv,* 29, 257–329.

Kristiansen, J. (1961). Sexual reproduction in *Mallomonas caudata. Botanisk Tidsskrift,* 57, 306–309.

– (1963). Sexual and asexual reproduction in *Kephyrion* and *Stenocalyx* (Chrysophyceae). *Botanisk Tidsskrift,* 59, 244–254.

– (1965). The occurrence and ecology of *Chrysolokos planctonicus,* a chrysomonad with sexual reproduction. *Botanisk Tidsskrift,* 61, 89–105.

Lehman, J. T. (1976a). Ecological and nutritional studies on *Dinobryon* Ehrenb.: seasonal periodicity and the phosphate toxicity problem. *Limnology and Oceanography,* 21, 646–658.

– (1976b). Photosynthetic capacity and luxury uptake of carbon during phosphate limitation in *Pediastrum duplex* (Chlorophyceae). *Journal of Phycology,* 12, 190–193.

Mack, B. (1951). Morphologische und entwicklungsgeschichtliche Untersuchungen an Chrysophyceen. *Österreichische Botanische Zeitung,* 98, 249–279.

– (1954). Untersuchungen an Chrysophyceen. 5–7. *Österreichische Botanische Zeitung,* 101, 64–73.

Margalef, R. (1958). Temporal succession and spatial heterogeneity in phytoplankton. In *Perspectives in Marine Biology,* ed. A. A. Buzzati-Traverso, pp. 323–349. Berkeley: University of California Press.

Meyer, R. L. (1965). *Morphology, Cytology, and Life History of Certain Chrysophyceae.* Ph.D. Thesis, University of Minnesota.

Munch, C. S. (1980). Fossil diatoms and scales of Chrysophyceae in the recent history of Hall Lake, Washington. *Freshwater Biology,* 10, 61–66.

Neff, R. J., & Neff, R. H. (1969). The biochemistry of amoebic encystment. *Symposia of the Society for Experimental Biology,* 23, 51–81.

Nygaard, G. (1956). Ancient and recent flora of diatoms and Chrysophyceae in Lake Gribsø. *Folia Limnologica Scandinavica,* 8, 32–94.

Pascher, A. (1924). Zur Homologisierung der Chrysomonadencysten mit den Endospore der Diatomeen. *Archiv für Prosistenkunde,* 48, 196–203.

– (1932). Über die Verbreitung endogener bzw. endoplasmatisch gebildeter Sporen bei den Algen. *Beiheft zum Botanischen Centralblatt,* 49, 293–308.

Pickett-Heaps, J. D. (1975). *Green Algae: Structure, Reproduction, and Evolution in Selected Genera.* Sinauer, Sunderland, Mass.

Prowazek, S. (1903). Flagellatenstudien. *Archiv für Protistenkunde,* 2, 195–212.

Ruttner, F. (1930). Das Plankton des Lunzer Untersees. *Internationale Revue der gesamten Hydrobiologie und Hydrographie,* 23, 1–304.

Sandgren, C. D. (1977). An ultrastructural investigation of statospore development in *Dinobryon cylindricum* Imhof (Chrysophyceae, Chrysophycophyta). *Journal of Phycology,* 13, (Suppl.), 59.

– (1978). *Resting Cysts of the Chrysophyceae: Their Induction, Development, and Strategic Significance in the Life History of Planktonic Species.* Ph.D. Thesis, University of Washington, Seattle.

– (1980a). An ultrastructural investigation of resting cyst formation in *Dinobryon cylindricum* Imhof (Chrysophycota). *Protistologica,* 16, 259–276.

– (1980b). Resting cyst formation in selected chrysophyte flagellates: an ultrastructural survey including a proposal for the phylogenetic significance of interspecific variations in the encystment process. *Protistologica,* 16, 289–303.

– (1981). Characteristics of sexual and asexual resting cyst (statospore) formation in *Dinobryon cylindricum* Imhof (Chrysophyceae, Chrysophyta). *Journal of Phycology,* 17, 199–210.

Scherffel, A. (1911). Beitrag zur Kenntnis der Chrysomonadineen. *Archiv für Protistenkunde,* 22, 299–344.

Sheath, R. G., Hellebust, J. A., & Sawa, T. (1975). The statospore of *Dinobryon divergens* Imhof: formation and germination in a sub-arctic lake. *Journal of Phycology,* 11, 131–138.

Skuja, H. (1950). Körperbau und Reproduktion bei *Dinobryon borgei* Lemm. *Svensk Botanisk Tidskrift,* 44, 96–107.

Smith, G. M. (1950). *Freshwater Algae of the United States.* New York: McGraw–Hill.

Smol, J. P. (1980). Fossil Synuracean (Chrysophyceae) scales in lake sediments: a new group of paleoindicators. *Canadian Journal of Botany,* 58, 458–65.

Stewart, K. D., & Mattox, K. R. (1975). Comparative cytology, evolution and classification of the green algae with some consideration of the origin of other organisms with chlorophylls *a* and *b*. *Botanical Review,* 41, 104–135.

Tippet, R. (1964). An investigation into the nature of the layering of deepwater sediments in two Eastern Ontario Lakes. *Canadian Journal of Botany,* 42, 1693–1709.

Vetter, H. (1937). Limnologische Untersuchungen über das Phytoplankton und seine Beziehungen zur Ernährung der Zooplanktons im Schleinsee bei Langenargen am Bodensee. *Internationale Revue der gesamten Hydrobiologie und Hydrographie,* 34, 499–561.

Wujek, D. E. (1969). Ultrastructure of flagellated Chrysophytes. I. *Dinobryon. Cytologia,* 34, 71–79.

3

Diatom resting spores: significance and strategies

PAUL E. HARGRAVES AND FRED W. FRENCH

Graduate School of Oceanography
University of Rhode Island
Kingston, RI 02881

3.1 Diversity of structure

Diatom resting spores, or hypnospores, are heavily silicified stages in the life cycles of marine centric diatoms and a few freshwater and pennate diatoms. In some genera, resting spores superficially resemble the parent vegetative cell (Fig. 1), whereas in others, spores and vegetative cells are morphologically quite different (Fig. 2; and Hargraves, 1976, 1979).

In Ross et al. (1979), spores are described as exogenous, semi-endogenous, or endogenous depending on how mature spores appear relative to the valves of the parent vegetative cell (Fig. 3). Spore valves are termed primary (first formed) or secondary and often differ structurally from each other. Girdle bands may or may not be present depending on genus (present: *Detonula, Thalassiosira, Eucampia, Odontella;* absent: *Bacteriastrum, Chaetoceros, Leptocylindrus, Stephanopyxis;* unknown or variable: *Cerataulina, Rhizosolenia, Bacterosira, Ditylum*). Likewise, the presence of pores or areolae passing through spore valves vary according to the genus involved, but as a general rule, if girdle bands are present, so are pores and/or areolae (exception: *Stephanopyxis*).

Von Stosch and Fecher (1979) recently described a morphologically distinct resting spore for the freshwater pennate *Eunotia soleirolii* (Kützing) Rabenhorst but list only seven pennates known to form resting spores, five of them freshwater and two marine. Certainly, the occurrence of morphologically distinct resting spores is rare among the pennate diatoms.

Studies on physiologically resting cells that do not differ morphologically from normal vegetative cells are not discussed in this chapter, al-

We gratefully acknowledge the technical contributions of the following persons: P. Boyd, J. Bullinger, C.-J. Chen, P. Degidio, D. French, J. Goldberg, P. Johnson, H. Rines, E. Rodehorst, D. Scales, and D. Stockwell. Parts of this work were supported through NSF OCE74-02293 and OCE76-82280.

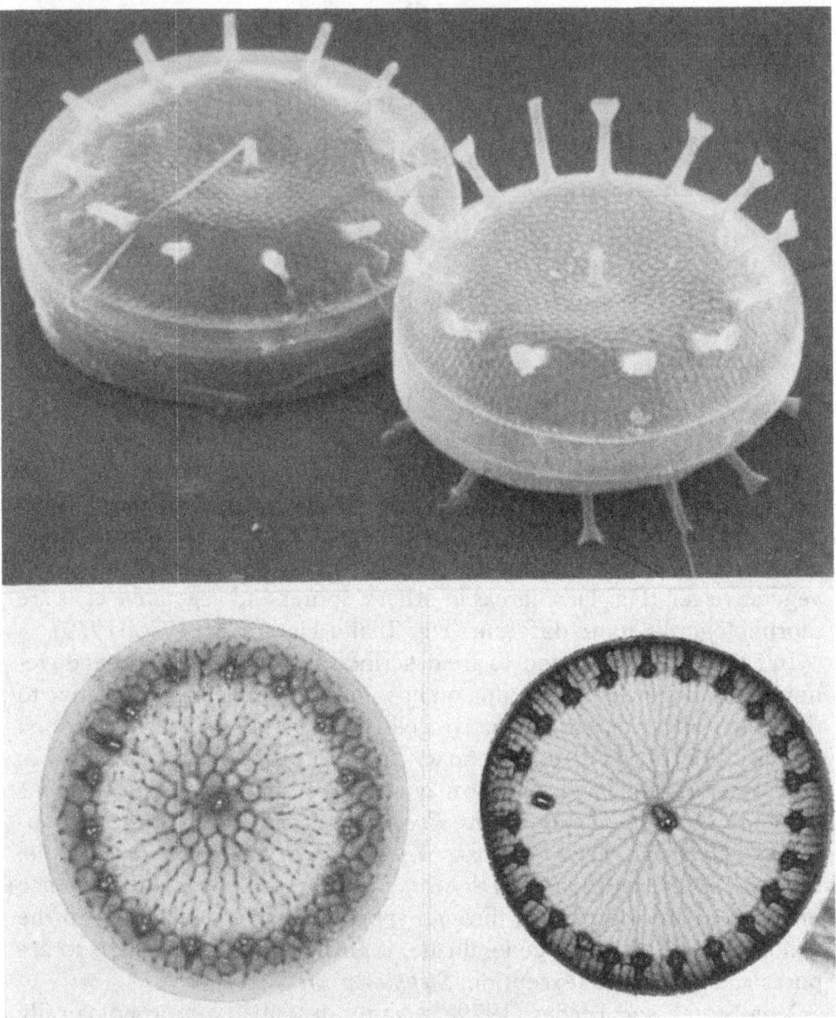

Fig. 1. Resting spores that resemble corresponding vegetative cells. Upper, *Thalassiosira nordenskioeldii* Cleve (SEM, ×1500); lower, *Detonula confervacea* (Cleve) Gran (TEM, ×5,500). Left, spore; right, vegetative cell.

though the possibility exists that many species may form such cells (see Anderson, 1975, 1976; and survival records for diatoms in Davis, 1972).

3.2 Phylogeny

Simonsen (1979) presents a scheme as his latest distillation of ideas on diatom phylogeny, which is adapted here to illustrate the occurrence of spores (Fig. 4). Whether occurrence of spores is common (C) or occa-

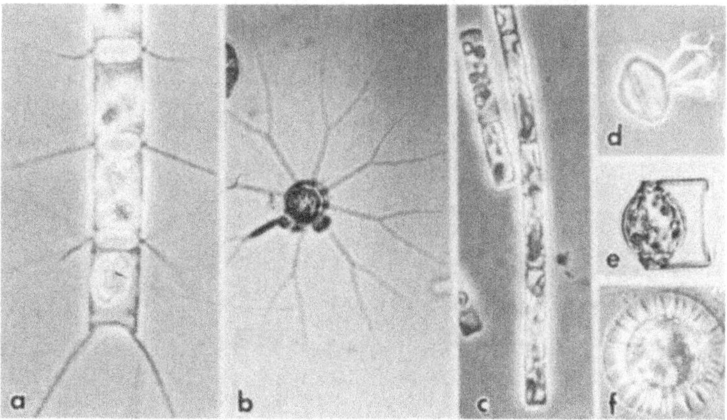

Fig. 2. Resting spores that differ from corresponding vegetative cells. *Chaetoceros diadema:* a, vegetative; d, spore. *Bacteriastrum delicatulum:* b, vegetative; e, spore. *Leptocylindrus danicus:* c, vegetative; f, spore. (Light photomicrographs: a, b, c = ×400; d, e, f = ×500.)

sional (O) within the family is noted for families having genera that form resting spores.

Simonsen accepts the argument of Ross and Sims (1973) that resting spores indicate neritic species, but he also states that, based on phylogeny, "species capable of forming resting spores are generally on the lower ranks of evolution," thereby concluding that "the Bacillariophyceae originated in inshore marine habitats."

This phylogeny is highly speculative. Circled areas and asterisks in Figure 4 indicate areas of maximum uncertainty. Stippling indicates our view of an approximate geologic time after which newly evolved families have either lost or never possessed the ability to form spores, for example, no families that evolved during or after the time indicated by the stippled area have extant members that form morphologically distinct spores; this is also true for some families that evolved before this time. The exact time represented by the stippled area in Fig. 4 is uncertain; probably it was sometime in the Miocene.

It is likely that "spore genera" (i.e., fossil diatom genera that may be spores) are extinct members of families that evolved before the time indicated by the stippled area but which are not known to form spores today (see, for example, Jousé, 1978).

3.3 Biogeographical considerations

In general, resting spores are found more often in coastal waters than in oceanic regions. The strict characterization of species as "oceanic" or "neritic" is difficult and in many cases is subjective. Many spore-forming species that are nominally coastal may be found in abundance and occa-

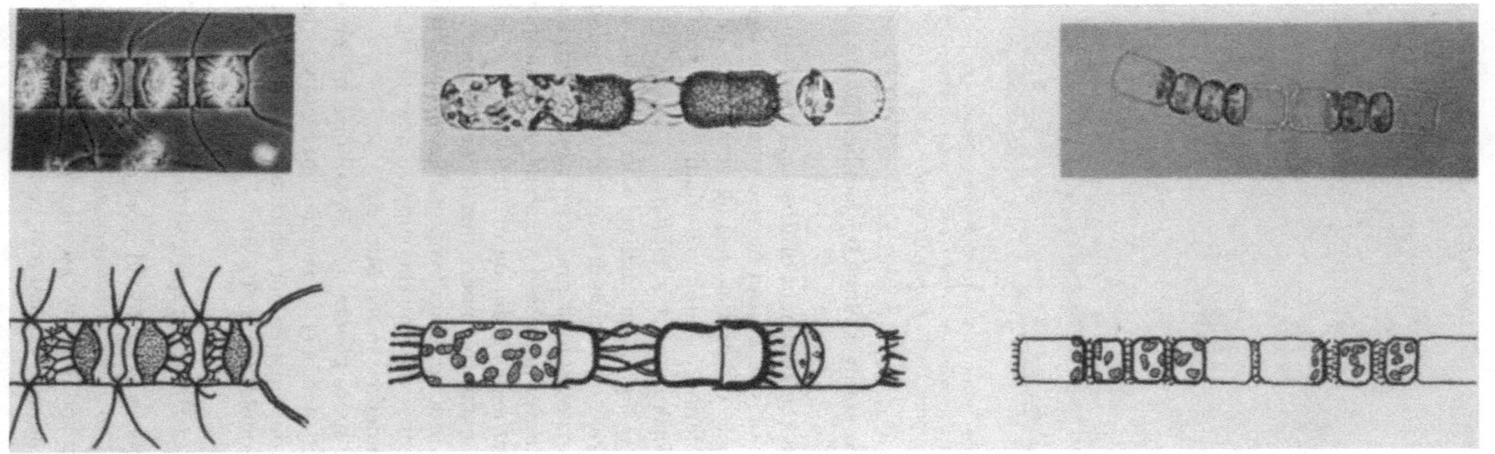

Fig. 3. Modes of spore formation (after Ross et al., 1979). Left, endogenous; center, semi-endogenous; right, exogenous.

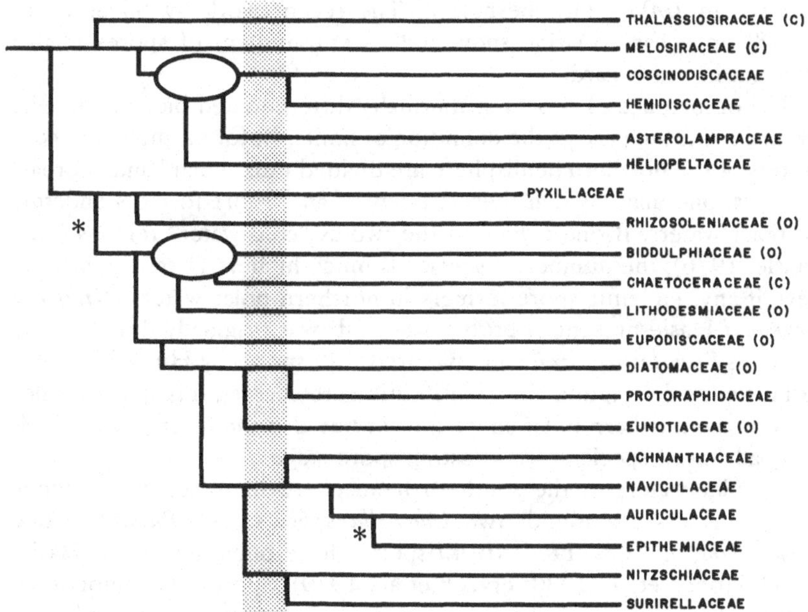

Fig. 4. Tentative phylogeny of diatom families (modified from Simonsen, 1979). Asterisks and circles indicate maximum uncertainty. For further discussion, see text.

sionally produce resting spores in oceanic waters (personal observation; Gaarder, 1951; Paasche, 1960, 1961; Simonsen, 1974). Such species include *Chaetoceros anastomosans* Grunow, *Chaetoceros pavillardii* Ikari, *Chaetoceros lorenzianum* Grunow, *Chaetoceros teres* Cleve, *Leptocylindrus danicus* Cleve, and others. Conversely, nominal oceanic species with the ability to form resting spores may be found in coastal waters (e.g., *Bacteriastrum delicatulum* Cleve). The extent to which the occurrence of "coastal" species in "oceanic" waters is a function of water mass entrainment, as is now well documented for the Gulf Stream system, cannot be accurately evaluated at this point. Nor is the state of our knowledge sufficient to say how extensive infraspecific variation in the ability to form spores may be, although there is some evidence for *L. danicus* (see section 3.4).

There is a long-standing assumption that the presence of resting spores in itself characterizes a species as neritic, the argument being that rapidly sinking spores would be a competitive disadvantage to an oceanic species. If the applicability of our data (see section 3.6; Table 3) on relative sinking rates of vegetative cells and spores can be shown to be a common characteristic of diatom spores, this assumption may be incorrect. Spores of an oceanic species need only sink to the pycnocline (which would function as a pseudobottom) until resuspension and germination take place. This idea is not totally new, having been obliquely suggested by

Karsten in 1905 (Schwebesporen). The recent work by Silver et al. (1978) on oceanic marine snow and its entrainment of spores is also relevant in this regard.

The occurrence of spores in thermally differentiated biogeographical areas depends in part on the definition of limits. If, for example, the cold waters of the northern hemisphere are divided into "polar" and "boreal" regimes, one may say that there are few, if any, spore-formers endemic to polar waters. If one combines the two as "cold-water" regions (e.g., Hasle, 1976), the number of species is much higher. There seem to be few, if any, endemic spore-formers in northern polar waters. *Nitzschia grunowii* Hasle presents a problem here. It was originally described by Cleve as *Fragilaria oceanica* and pictured with spores by Hustedt (1959). If Hustedt's description and identification are correct, this species is not only a true northern polar spore-former but also the most phylogenetically advanced species with a resting spore stage.

In polar waters of the southern hemisphere likewise, few endemic spore formers are found: two *Odontella* species, one *Eucampia*, one *Chaetoceros*, and one *Thalassiosira* species have been described (Hasle, 1969; Hoban et al., 1980; Fryxell et al., 1979). Spores of cosmopolitan species that occur in southern polar waters are extremely rare. In boreal and temperate regions of both hemispheres, the occurrence of spore-formers is at its peak, with dozens of species in several genera, particularly *Chaetoceros*, *Thalassiosira*, *Rhizosolenia*, and *Stephanopyxis*. Many of these species have extended distributions into colder and warmer waters.

Tropical regions are comparatively depauperate in the occurrence of resting spores. Almost without exception, resting spores in tropical waters are formed by species with a much wider (i.e., cosmopolitan) distribution. One possible exception is *Chaetoceros melchersianum* Margalef, a species of questionable identity. Resting spores appear to be a common phenomenon in upwelling areas in the tropics, but this is essentially an "island" of temperate water.

Causes for the comparative lack of resting spores in polar and tropical regions is a fruitful area for investigation. Attributing this comparative lack to temperature differences alone is too simplistic an answer. If nutrient stress is the most important trigger in spore induction (see section 3.5), perhaps the situation in polar waters is partly explained: Nutrients (at least in southern polar waters) are probably not limiting. But what of the tropics? Perhaps the key to spore formation is fluctuation in stress conditions. Compared to polar and temperate regions, the tropics are characterized (except for near-shore and upwelling areas) by relative uniformity of environment. Likewise for spore germination, fluctuation in stress might be important. If germination of spores occurs as environmental stress is released, as discussed below, the lack of release from environmental stress in the tropics would eventually select against spore-forming species, except in cosmopolitan species that are regularly reintroduced.

Resting spores in freshwater diatoms are rather uncommon both in natural frequency and number of species in which formation occurs. Predominantly marine genera with spore-forming freshwater representatives include *Chaetoceros* (two species) and *Rhizosolenia* (two species). *Melosira* can be mentioned here as well, although Crawford (1979) suggests that four separate genera might be established among the many species of this unwieldy genus. *Melosira italica* (Ehrenberg) Kützing can form spores, which may have a role in the annual cycle of this species in some lakes (Lund, 1954). It is noteworthy that most of the pennate diatoms that form spores are restricted to freshwater habitats: *Achnanthes hungarica* Grunow; *Diatoma anceps* (Ehrenberg) Kirchner; *Meridion circulare* (Greville) Agardh; and at least four species of *Eunotia* (von Stosch & Fecher, 1979). Of these pennate species, only in *E. soleirolii* has the distinction between "internal thecae" and a true resting spore been unequivocally demonstrated. The unusual diatom *Acanthoceras zachariasii* (Brun) Simonsen (Simonsen, 1979) also forms spores, but little is known of the process. Most of the freshwater species forming spores are widely distributed in temperate latitudes.

3.4 Life cycle patterns

Resting spores are normally vegetative in mode of formation (von Stosch & Drebes, 1964; Drebes, 1966; von Stosch et al., 1973). However, in *L. danicus* (Fig. 2), the resting spore is the result of sexual processes, forming inside the auxospore (French, 1980), and this may also be true of *Cerataulina pelagica* (Cleve) Hendey (Saunders, 1968).

The ability to form resting spores may be limited to certain clones in some species. In *L. danicus*, for example, spores are only rarely seen in nature. In laboratory studies, only 3 out of more than 50 clones we isolated from Narragansett Bay have been capable of forming resting spores (personal observation). Because spores of many other species, such as *C. pelagica, Chaetoceros pseudocurvisetum* Mangin, and *Porosira glacialis* (Grunow) Jörgensen are also rare in nature, the hypothesis of clonal variation in spore-forming ability clearly needs to be tested in a wide variety of species. If there is clonal variation in the ability to form resting spores, this also might explain the relative lack of spore formation in cosmopolitan species occurring in tropical waters.

The vegetative cell-size range within which resting spores may form varies from species to species, as illustrated in Fig. 5, but generally exceeds the size range within which sexuality can occur. It should be pointed out that few species have been examined in this regard.

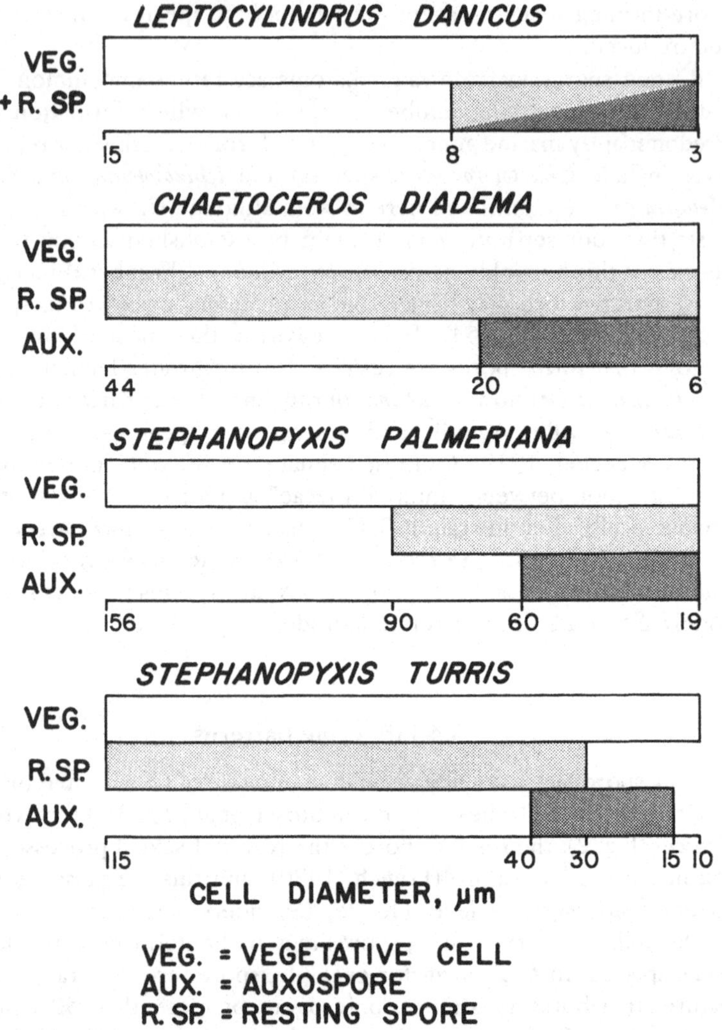

Fig. 5. Comparison of vegetative cell, resting spore, and auxospore size ranges for four diatoms.

3.5 Formation

As outlined by von Stosch and Kowallik (1969), a mitotic division accompanies the formation of each resting spore valve. Spore formation may be the result of an unequal cytokinesis, as in the presumably primitive genus *Stephanopyxis* (von Stosch & Drebes, 1964; Drebes, 1966) or of two "acytokinetic mitoses," as in *Chaetoceros didymum* Ehrenberg (von Stosch et al., 1973). Following primary valve formation, a coalescing of cell organelles takes place before secondary valve deposition (Fig. 6).

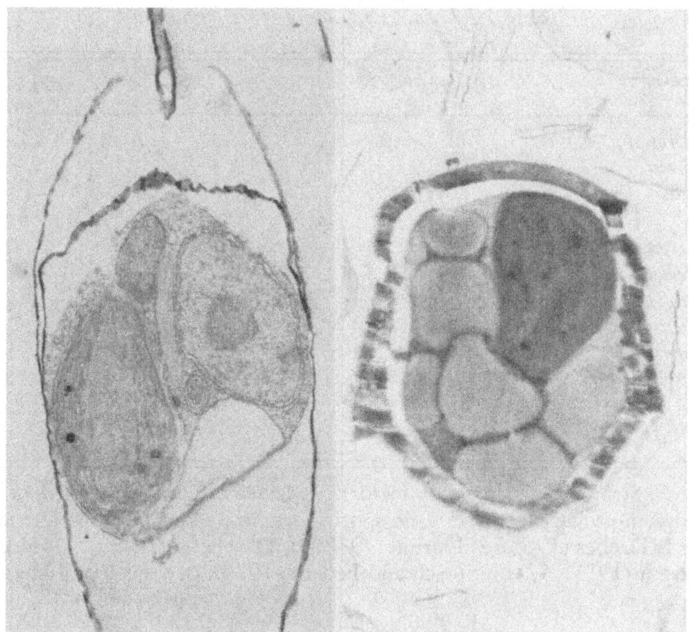

Fig. 6. Organelle structure during spore formation in *Chaetoceros sociale* f. *radians*. Left: after first formation of primary valve, enclosed in vegetative cell. Right: free, fully formed spore (separation of cytoplasm from theca due to stresses in sectioning). Note compaction of organelles and elimination of vacuolar space (TEM, both ×14,000.)

The process has been strikingly demonstrated cinematically by Drebes (1975).

Numerous environmental variables have been implicated as contributory to inducing resting spore formation. These are summarized in Table 1 for the various studies that have looked at this problem specifically. Nitrogen deficiency is a causative factor in every case, often as the only factor that will trigger formation. We have been unable to induce some known spore-formers to encyst by any means (e.g., *C. pelagica*) and have succeeded with only certain clones of other species (*L. danicus*). As suggested by Paasche (1961), the possibility of geographical races existing within a species, some capable of spore formation, needs further investigation (see also Antia, 1976).

3.6 Longevity and resistance

Table 2 presents data from the available literature plus personal observations on survival of diatom resting spores. All spores were held in darkness unless otherwise noted. Data on survival of non–spore-forming diatoms are widely scattered in the literature, but recent publications

Table 1. *Variables inducing spore formation*

	Source	N	P	Si	Fe	T	pH	Light
Chaetoceros spp. (3)	4							(−)
C. diadema	6	++						
C. sociale	6	++						
C. teres	6	++						
D. confervacea	2	++				(+)		
D. confervacea,								
E. soleirolii	5	+	+	+	++	++	+	
L. danicus	3	++						
L. danicus	6	++				(+)		
S. palmeriana	1	+	++					
S. turris	6	++						
T. nordenskioeldii	2	++				(+)		

Note: ++, Strong effect (primary factor); +, lesser effect; (+), has effect when coupled with primary factor; (−), no spores form in darkness in complete media.
Sources: 1, Drebes (1966); 2, Durbin (1978); 3, Davis et al. (1980); 4, Hargraves and French (1975); 5, von Stosch and Fecher (1979); 6, French and Hargraves (1980).

Table 2. *Resting spore survival time (in days)*

		Temperature (°C)					
Species	Source	0	2–3	5–6	10	15	20
Chaetoceros diadema	2		140*				
C. diadema	4 (D)		291*	291*	291*	291*	
	(L)		291*	291*	291*	95	
C. diadema	6		645				
C. didymum	6		645				
C. sociale f. *radians*	2					140*	
C. vanheurckii	6		645				
D. confervacea	1	576		220	85	42	<7
D. confervacea	2		140*				
E. soleirolii	5	58	58	400		400	
L. danicus	3		214		214		
L. danicus	4 (D)				401	191	217
	(L)				191		
S. turris	2				112		
T. nordenskioeldii	1	576		220	105	42	<7

Note: *, Longest survival time tested; (D), held in dark; (L), held in dim light (20 $\mu E/m^2/sec$), 12:12-hr photoperiod.
Source: 1, Durbin (1978); 2, Hargraves and French (1975); 3, Davis et al. (1980); 4, personal observation; 5, von Stosch and Fecher (1979); 6, Hollibaugh et al. (1981).

dealing in part with survival of vegetative diatom cells are Antia (1976), Davis (1972), and Smayda and Mitchell-Innes (1974).

It should be pointed out that these are simply the longest times demonstrated for these species to date, making these conservative estimates. No one has rigorously tested any species on a weekly basis over a wide interval of temperatures to produce a precise data set on survival time; Durbin's (1978) study is the most complete.

Several trends emerge. First, spores tend to survive longer at colder temperatures, with the freshwater pennate *E. soleirolii* the exception. Second, the spores of boreal species do not appear to tolerate temperatures higher than do their vegetative cells, as pointed out by Durbin (1978), whereas spores of temperate species (e.g., *L. danicus*) survive relatively long periods at elevated temperatures. Third, the role of spores appears to be tied at the most to an annual cycle, as no centric spore has survived 2 years.

Several other observations are of interest. Darkness appears to prolong, but is not necessary, for spore survival, as spores survive substantial periods in low light (personal observation, see Table 2). Davis et al. (1980), during observations in the CEPEX Controlled Environmental Ecosystems (CEEs), found that *L. danicus* spores collected directly after formation could germinate 2 weeks later but that spores allowed to sit in bottom sediments of the CEEs for as little as 5 days failed to germinate. They attributed this to conditions in these sediments, such as anoxia and high concentrations of ammonia, hydrogen sulfide, and organic matter. This suggests bottom condition limits under which spores could be expected to survive.

In cores taken from Narragansett Bay and Block Island Sound, we (Hargraves & French, 1975) found spores in sediments in low numbers and only during and shortly after the planktonic occurrence of the species involved. We concluded that spores deposited inshore did not survive there to seed the following year's population, with conditions for survival in deeper water offshore unknown. Spores of boreal species cannot survive summer in shallow temperate waters, because they have no more thermal resistance than vegetative cells. Our limited success in finding spores in sediments could be due in part to our observational techniques but may also be influenced by the variables mentioned by Davis et al. (1980) on anoxia. No laboratory experiments have been performed to test spores' ability to withstand anoxia, high NH_4^+, high H_2S, or high concentrations of organic material, although experimental techniques are available to do so, and spores do respire (French & Hargraves, 1980; see section 3.8).

Resting spores generally survive periods of air-drying better than vegetative cells do (Hargraves & French, 1975). This would enhance the ability for aerial transport, known to take place in diatoms, or for other

Table 3. *Sinking rates of vegetative cells and spores (in m · day^{-1})*

Species	Vegetative cells	Vegetative chains	Resting spores
C. teres	$\bar{x}$ = 4.8	$\bar{x}$ = 5.7	$\bar{x}$ = 7.6
	N = 404	N = 179	N = 64
	R = 1.3–10	R = 1.6–12	R = 2.0–16
D. confervacea	$\bar{x}$ = 1.9	$\bar{x}$ = 1.9	$\bar{x}$ = 1.8
	N = 51	N = 55	N = 12
	R = 1.0–3.2	R = 1.0–4.0	R = 1.3–5.3
S. turris	$\bar{x}$ = 3.3	$\bar{x}$ = 7.1	$\bar{x}$ = 3.2
(*small*)	N = 164	N = 70	N = 53
	R = 1.6–8.0	R = 3.2–12	R = 1.1–8.0
S. turris	$\bar{x}$ = 2.6		$\bar{x}$ = 6.5
(*large*)	N = 44		N = 71
	R = 1.1–5.3		R = 1.0–16.0
L. danicus	$\bar{x}$ = 1.43		$\bar{x}$ = 2.06
	s = 0.96		s = 1.42
	N = 2181		N = 192

Note: $\bar{x}$, Mean; s, standard deviation; N, number of observations; R, range.
Source: French and Hargraves (1980).

means of transport, for example, mud on the feet of wading birds. To our knowledge, however, no observations exist that show spores actually have a role in such transport.

Data on sinking rates of spores and vegetative cells of the same species are mostly lacking, with Table 3 (French & Hargraves, 1980) the only data of which we are aware. Spores sink somewhat faster than vegetative cells but the difference is not marked. The question should be asked as to whether the sole role of spores is to sink to the benthos. This does not appear to be the case in Narragansett Bay (Hargraves & French, 1975), but circumstantial evidence suggests that accumulation in the benthos exists in other geographic locations (Davis et al., 1980; Garrison, 1981). Silver et al. (1978) observed resting spores in marine snow (organic aggregates), often representing 100% of the resting spores observed in the water column. The same thing is often true in culture, where spores are found in clumps clinging to any debris. Because spores can germinate in as little as several days after formation (see section 3.7), one hypothesis is that some spores never sink to the bottom at all, their spines or collars causing their incorporation in marine snow, where they will germinate when again exposed to adequate nutrients. Marine snow is enriched in nutrients, particularly nitrogen (Shanks & Trent, 1979; Alldredge, 1979). This hypothesis need not replace the traditional idea of spores sinking and persisting in sediments over longer periods but offers

Table 4. *Factors inducing spore excystment*

Species	Source	Nutrients	Light	Temperature
B. hyalinum	6	?	+	(+)
C. diadema	4	++	+	
D. conferacea	2	++	+	
E. soleirolii	5	?		++
L. danicus	3	++	+	(+)
L. danicus	4	++	+	
R. setigera	5	+		(+)
S. palmeriana	1	++		
T. nordenskioeldii	2	++	+	

Note: Symbols as in Table 1.
Sources: 1, Drebes (1966); 2, Durbin (1978); 3, Davis et al. (1980); 4, Hargraves and French (1975); 5, von Stosch and Fecher (1979); 6, Drebes (1972).

an interesting additional direction for thought and experimentation: Germinating spores may indeed be responsible for sudden appearances of species, but introduction of spores may be from a pelagic environment rather than a benthic one.

3.7 Excystment

Resting spore germination varies in detail from species to species. In presumably primitive *Stephanopyxis*, the spore elongates, and mitosis is followed by new vegetative valve formation, with the spore valves retained by both new vegetative cells (von Stosch & Drebes, 1964). In *C. didymum*, two new vegetative valves are formed within the spore valves, each following an acytokinetic mitosis, and the old spore valves are cast off (von Stosch et al., 1973). The same is true of *L. danicus* (French, 1980). In the pennate *E. soleirolii* (von Stosch & Fecher, 1979), a complex excystment produces two large viable cells and two smaller cells, each bearing a spore valve. These small cells usually die, making this excystment energetically inefficient and quite different from the process in centric species.

Factors inducing excystment are listed in Table 4. The bias toward the importance of nutrients may be our own, based on our observations that spores can remain for months in nitrogen-depleted medium in light without excysting but will excyst rapidly when nutrients are added. Adequate light is also needed, however, as spores will not excyst in darkness in complete media. The minimum light intensity needed for excystment is probably just in excess of compensation intensity, as *C. diadema* will excyst at 10°C in 20 $\mu E/m^2/sec$ (personal observation). Generally the higher the light level (within physiological tolerances), the more rapid the excystment.

Temperature plays an ambiguous role. Generally, excystment takes place more quickly at higher temperatures. One question concerns the idea of dormancy: Do resting spores require a period of cold (and dark) before excysting? Unquestionably, this is required by spores of the freshwater pennate (*E. soleirolii*; von Stosch & Fecher, 1979). However, among marine centric diatoms, there instead appears to be a short refractory period during which spores will not germinate, usually several days to several weeks (e.g., *C. diadema,* 3 days; *L. danicus,* 5 to 6 days). After this period, spores will germinate in renewed medium, even if held during the interim at the temperature of formation and the same diurnal light:dark cycle. Our conclusion, in agreement with von Stosch & Fecher (1979), is that centric diatoms' resting spores do not pass through a period of true dormancy in the sense that dinoflagellate cysts do (Wall, 1975; Dale, this volume).

3.8 Physiology

Physiological measurements on spore cultures have been made by French and Hargraves (1980) and Hollibaugh et al. (1981). Compared with vegetative cells, spores (1) have a higher C/N ratio, (2) contain more carbon and chlorophyll per cell if darkened, (3) have much lower carbon-specific respiration if held in darkness, and (4) can continue to photosynthesize after at least a month in darkness (French & Hargraves, 1980). Using collections from *in situ* large-volume plastic enclosures, Hollibaugh et al. (1981) examined some physiological characteristics of the spores of three *Chaetoceros* species. When removed from the dark and placed under conditions suitable for germination, spores immediately began to fix carbon but did not begin to take up nutrients or synthesize chlorophyll for 24 hr or more.

Spores appear specialized to persist in darkness or low light, especially in nutrient-depleted media, but also in high-nutrient media, if darkness or subexcystment light intensities are maintained. Spores definitely outsurvive vegetative cells in complete darkness. Spore organelles do not appear to break down, although they are more condensed than they are in the vegetative cell due to the mode of spore formation, and spores become progressively more packed with what presumably are vesicles of a storage product (see Fig. 6). The nature of this storage product is under investigation.

Continuing high photosynthetic ability coupled with low respiration means that occasional resuspension of a spore into the euphotic zone for short periods could "recharge" it like a battery (French & Hargraves, 1980), thus permitting extended survival without germination.

Table 5. *Resistance of spores to grazing*

Species	No. of fecal pellets produced	% fecal pellets with spores	% germination
C. diadema	4	100	0
C. laciniosum	16	94	0
C. sociale f. *radians* (clone B7)	24	100	50
C. sociale f. *radians* (clone P1OH)	21	76	42
C. sociale f. *radians* (clone B5)	1	100	100
C. sociale f. *radians* (clone CHSC3)	7	100	0
C. sociale f. *radians* (clone CHSC4)	5	100	40
C. teres	8	100	0
Chaetoceros sp.	3	100	0
D. confervacea (clone DNH)	18	100	55
D. confervacea (clone Det-1)	14	100	0
D. confervacea (clone Dcon)	2	100	0
L. danicus	10	60	0
S. turris	19	47	0

3.9 Grazing interactions

A series of papers by Porter (1973, 1975, 1976) has demonstrated the importance of phytoplankton survival during and following ingestion by grazers. Zooplankton grazing suppressed or did not affect the numbers of all groups of phytoplankton, except gelatinous green algae (e.g., *Sphaerocystis schroeteri* Chodat). This group increased because they could survive passage through zooplankton guts and "are found growing profusely from copepod fecal pellets in the lake" (Porter, 1973), even taking up nutrients from the fecal pellet, which further enhanced growth. This sequence may "drive seasonal succession." Diatoms either did not physically survive gut passage or did not reproduce thereafter.

Similar experiments performed with a number of marine diatom cultures yielded the results in Table 5. Only cultures containing spores ever survived gut passage and grew again. Cultures showing high survival rates contained a high percentage of spores, and the converse is also true. Thus this table is probably an underestimate of the number of species capable of surviving grazer ingestion, since fecal pellets that were produced by zooplankton feeding on cultures with very few spores may have contained no spores. Figure 7 is a thin section through the gut of a copepod, *Acartia tonsa* Dana, fed a mixture of *C. sociale* f. *radians* vegetative cells and resting spores. The spore is intact whereas only fragments of the less heavily silicified vegetative cell frustules are present.

Incorporation into fecal pellets would speed passage to the sediments, and surviving ingestion could be important during periods of heavy graz-

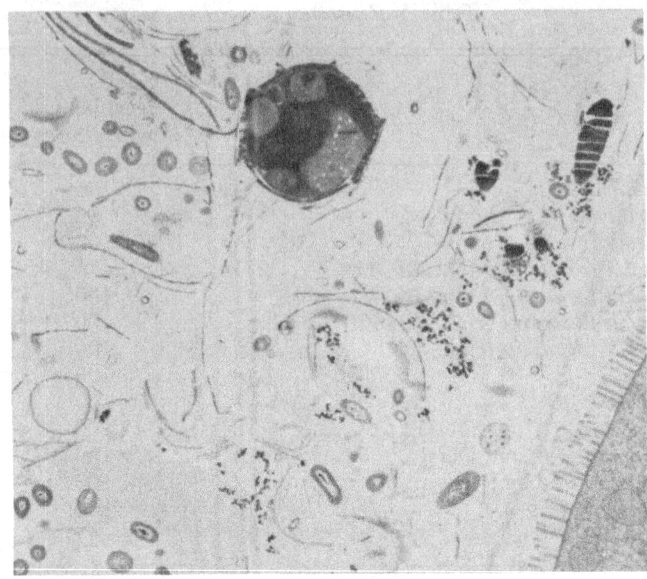

Fig. 7. Intact spore of *Chaetoceros sociale* f. *radians* in hindgut of copepod *Acartia tonsa*. Gut wall, lower right. Note numerous vegetative cell fragments. (TEM, ×4000.)

ing. Since fecal pellets are also found in marine snow (Silver et al., 1978), spores incorporated in pellets resulting from grazed marine snow could absorb nutrients and germinate shortly after formation, but perhaps in a more favorable external nutrient environment that that which induced their formation. However, the advantage of incorporation might be offset by increased sinking rates of these aggregates (Shanks & Trent, 1980).

3.10 Conclusions

Examining all the information currently available on diatom resting spores, the idea emerges that as a whole they function differently from similar stages in other algal classes. Very briefly, encystment and excystment are controlled by a nutrient–light combination (temperature is not a first-order factor), spores probably do not persist in nature for as much as 2 years; no true dormant state exists, and spores can excyst in as little as a few days.

If Simonsen is correct and the resting spore is a primitive characteristic, it may seem somewhat surprising that spores have not been replaced in all species by physiologically resting cells. Our observations on *L. danicus* indicate that resting spore formation "costs" one or more cell divisions in terms of final cell yield. That is, for a given amount of limiting nutrient, a culture can divide one more time if growing under conditions that preclude spore formation. If this is true for other species, it would then appear selectively advantageous for a species to dispense with spore formation entirely, go through the maximum number of divisions possible, and produce twice as many potential resting cells as it could spores.

More recently evolved genera (and species?) may have done just that, eliminating resting spore formation: Because they are morphologically undifferentiated, they would not easily be recognized in natural populations. The ubiquitous and abundant *Skeletonema costatum* (Greville) Cleve is one likely candidate for forming physiological resting cells (see discussion in Oppenheimer, 1966, pp. 187–188). Why has this not commonly happened in older genera as well? Spore formation is either genetically linked to other indispensable aspects of cell metabolism and therefore cannot be abandoned or there are as yet unknown selective advantages to retention of resting spores as a strategy in older genera.

In looking for such an advantage, the two morphological features that are most apparent are the heavily silicified spore frustule and the compaction of cellular contents in the spore. The heavy frustule, by increasing sinking rate at least incrementally, gets the spore more quickly to new nutrients or removes it from dangerously high light in the absence of such nutrients. Perhaps compaction of organelles has a similar protective effect in high light. Spores may serve a different function than is served by resting cells, at least initially, since the latter appear to form in darkness or low light.

If one keeps in mind that spore-forming diatom species are found mostly in temperate or boreal waters, a statement by Davis et al. (1980) may indicate the key advantage to spore formation: "because of the exponential nature of phytoplankton growth, the limiting nutrient is reduced from a considerable level to near zero during the last doubling of the phytoplankton at the end of a bloom (in one day or less for many diatoms)." They then present references for temperate zone winter–spring blooms often ending due to nutrient depletion rather than grazing or other factors. If we carry this view further, such a population, suddenly nutrient depleted and entrained in high light, might be totally destroyed due to photooxidative effects and metabolic imbalance, were it not for spore formation. A cursory examination of spore-forming ability shows that many spore-forming species are precisely those that are abundant in temperate winter–spring blooms. Spores can form quickly in these species under conditions (high light and low nutrients) in which physiological resting cells might not have a chance to form unless they were transported to areas conducive to resting cell formation (e.g., the benthos).

Our knowledge of the ecological and physiological aspects of resting spores is scant, especially in the area of comparing resting spores with resting cells. In addition to nutrient kinetics, a thorough knowledge of life cycles, clonal variation, and biogeographic distribution of spore-formers will be necessary before the role of resting spores will be thoroughly understood.

References

Alldredge, A. L. (1979). The chemical composition of macroscopic aggregates in two neritic seas. *Limnology and Oceanography*, **24**, 855–866.

Anderson, O. R. (1975). The ultrastructure and cytochemistry of resting cell formation in *Amphora coffaeformis* (Bacillariophyceae). *Journal of Phycology*, **11**, 272–281.

– (1976). Respiration and photosynthesis during resting cell formation in *Amphora coffaeformis* (Ag.) Kütz. *Limnology and Oceanography*, **21**, 452–456.

Anita, N. (1976). Effects of temperature on the darkness survival of marine microplanktonic algae. *Microbial Ecology*, **3**, 41–54.

Crawford, R. M. (1979). Taxonomy and frustular structure of the marine centric diatom *Paralia sulcata*. *Journal of Phycology*, **15**, 200–210.

Davis, C. O., Hollibaugh, J. T., Seibert, D. L. R., Thomas, W. H., & Harrison, P. J. (1980). Formation of resting spores by *Leptocylindrus danicus* (Bacillariophyceae) in a controlled experimental ecosystem. *Journal of Phycology*, **16**, 296–302.

Davis, J. S. (1972). Survival records in the algae, and the survival role of certain algae pigments, fat, and mucilaginous substances. *The Biologist*, **54**, 52–93.

Drebes, G. (1966). On the life history of the marine plankton diatom *Stephanopyxis palmeriana*. *Helgoländer Wissenschaftliche Meeresuntersuchungen*, **13**, 101–114.

– (1972). The life history of the centric diatom *Bacteriastrum hyalinum* Lauder. *Beihefte zur Nova Hedwigia*, **39**, 95–110.

– (1975). *Chaetoceros teres* (Centrales): Ungeschlechtliche Fortpflanzung, Zellteilung, Dauersporen (Film) *Encyclopedia Cinematographica*, Institut für Wissenschaftliche Film, Göttingen. 16 mm.

Durbin, E. (1978). Aspects of the biology of resting spores of *Thalassiosira nordenskioldii* and *Detonula confervacea*. *Marine Biology*, **45**, 31–37.

French, F. (1980). Diatom resting spores: a comparison of occurrence in the life cycles of *Chaetoceros diadema* (Ehr.) Gran and *Leptocylindrus danicus* Cl. *Journal of Phycology*, **16** (Suppl.), 11.

French, F., & Hargraves, P. E. (1980). Physiological characteristics of plankton diatom resting spores. *Marine Biology Letters*, **1**, 185-195.

Fryxell, G. A., Villareal, T. A., & Hoban, M. A. (1979). *Thalassiosira scotia*, sp. nov.: observations on a phytoplankton increase in early austral spring north of the Scotia Ridge. *Journal of Plankton Research*, **1**, 335–370.

Gaarder, K. R. (1951). Bacillariophyceae. *Report on the Scientific Results of the 'Michael Sars' North Atlantic Deep-Sea Expedition 1910*, **2**(2), 1–53.

Garrison, D. L. (1981). Monterey Bay phytoplankton. II. Resting spore cycles in coastal diatom populations. *Journal of Plankton Research*, **3**, 137–156.

Hargraves, P. E. (1976). Studies on marine plankton diatoms. II. Resting spore morphology. *Journal of Phycology*, **12**, 118–128.

– (1979). Studies on marine plankton diatoms. IV. Morphology of *Chaetoceros* resting spores. *Beihefte zur Nova Hedwigia,* **64**, 99–120.

Hargraves, P. E., & French, F. (1975). Observations on the survival of diatom resting spores. *Beihefte zur Nova Hedwigia,* **53**, 229–238.

Hasle, G. R. (1969). An analysis of the phytoplankton of the Pacific Southern Ocean. *Hvalrådets Skrifter,* **52**, 1–168.

– (1976). The biogeography of some marine plankton diatoms. *Deep-Sea Research,* **23**, 319–338.

Hoban, M. A., Fryxell, G. A., & Buck, K. R. (1980). Biddulphioid diatoms: resting spore formation in Antarctic *Eucampia* and *Odontella. Journal of Phycology,* **16**, 591–602.

Hollibaugh, J. T., Seibert, D. L. R., & Thomas, W. H. (1981). Observations on the survival and germination of resting spores of three *Chaetoceros* (Bacillariophyceae) species. *Journal of Phycology,* **17**, 1–9.

Hustedt, F. (1959). Die Kieselalgen. *Rabenhorst's Kryptogamen-Flora,* **7(2)**, 1–845.

Jousé, A. P. (1978). Diatom stratigraphy on the generic level. *Micropaleontology,* **24**, 316–326.

Karsten, G. (1905). Das Phytoplankton des Antarktischen Meeres nach dem Material der deutschen Tiefsee-Expedition, 1898-1899. *Deutsche Tiefsee Expedition,* **2(2)**, 1–136.

Lund, J. W. G. (1954). The seasonal cycle of the plankton diatom *Melosira italica* (Ehr.) Kütz. subsp. *subarctica* Müll. *Journal of Ecology,* **42**, 151–179.

Oppenheimer, C. H. (ed.) (1966). *Marine Biology,* Vol. II, *Phytoplankton.* New York: New York Academy of Sciences.

Paasche, E. (1960). Phytoplankton distribution in the Norwegian Sea in June, 1954, related to hydrography and compared with primary production data. *Reports on Norwegian Fishery and Marine Investigations,* **12(11)**, 1–77.

– (1961). Notes on phytoplankton from the Norwegian Sea. *Botanica Marina,* **2**, 197–214.

Porter, K. G. (1973). Selective grazing and differential digestion of algae by zooplankton. *Nature (London),* **244**, 179–180.

– (1975). Viable gut passage of gelatinous green algae ingested by *Daphnia. Internationale Vereinigung für Theoretische und Angewandte Limnologie,* **19**, 2840–2850.

– (1976). Enhancement of algal growth and productivity by grazing zooplankton. *Science,* **192**, 1332–1334.

Ross, R., Cox, E. J., Karayeva, N. I., Mann, D. B., Paddock, T. B. B., Simonsen, R., & Sims, P. (1979). An amended terminolgy for the siliceous components of the diatom cell. *Beihefte zur Nova Hedwigia,* **64**, 513–533.

Ross, R., & Sims, P. A. (1973). Observations on family and generic limits in the Centrales. *Beihefte zur Nova Hedwigia,* **45**, 97–132.

Saunders, R. P. (1968). *Cerataulina pelagica* (Cleve) Hendey. *Florida Board of Conservation Leaflet Ser.,* 1 (Pt. 2, No. 5), 1–11.

Shanks, A. L., & Trent, J. D. (1979). Marine snow: microscale nutrient patches. *Limnology and Oceanography,* **24**, 850–854.

– (1980). Marine snow: sinking rates and potential role in vertical flux. *Deep-Sea Research,* **27A**, 137–143.

Silver, M. W., Shanks, A. L., & Trent, J. D. (1978). Marine snow: microplankton habitat and source of small-scale patchiness in pelagic populations. *Science,* **201**, 371–373.

Simonsen, R. (1974). The diatom plankton of the Indian Ocean expedition of R/V Meteor, 1964-1965. *Meteor Forschungsergebnisse, Reihe* **D19**, 1–107.

– (1979). The diatom system: ideas on phylogeny. *Bacillaria,* **2,** 9–71.

Smayda, T. J., & Mitchell-Innes, B. (1974). Dark survival of autotrophic planktonic marine diatoms. *Marine Biology,* **25,** 195–202.

von Stosch, H. A., & Drebes, G. (1964). Entwicklungsgeschichtliche Untersuchungen an zentrischen Diatomeen. IV. Die Planktondiatomee *Stephanopyxis turris,* ihre Behandlung und Entwicklungsgeschichte. *Helgoländer Wissenschaftliche Meeresuntersuchungen,* **11,** 209–257.

von Stosch, H. A., & Fecher, K. (1979). "Internal thecae" of *Eunotia soleirolii* (Bacillariophyceae): development, structure and function as resting spores. *Journal of Phycology,* **15,** 233–243.

von Stosch, H. A., & Kowallik, K. (1969). Der von Geitler aufgestellte Satz über die Notwendigkeit einer Mitose für jede Schalenbildung von Diatomeen. Beobachtungen über die Reichweite und Überlegungen zu seiner zellemechanischen Bedeutung. *Österreichische Botanische Zeitschrift,* **116,** 454–474.

von Stosch, H. A., Theil, G., & Kowallik, K. (1973). Entwicklungsgeschichtliche Untersuchungen an zentrischen Diatomeen. V. Bau und Lebenszyclus von *Chaetoceros didymum* mit Beobachtunger über einige andere Arten der Gattung. *Helgoländer Wissenschaftliche Meeresuntersuchungen,* **25,** 384–445.

Wall, D. (1975). Taxonomy and cysts of red-tide dinoflagellates. In *First International Conference on Toxic Dinoflagellate Blooms,* ed. V. R. LoCicero, pp. 249–255. Wakefield: Massachusetts Science and Technology Foundation.

4

Dinoflagellate resting cysts: "benthic plankton"

BARRIE DALE
Department of Geology
University of Oslo
P. B. 1047 Blindern
Oslo 3, Norway

4.1 Introduction

Dinoflagellates are best known to biologists as one of the major groups of phytoplankton, important for their contribution to primary production and notorious for causing "red tides." They are an extremely varied group of eucaryotic organisms of both plant and animal affinities. Benthic, parasitic, and symbiotic forms are known, but biologists have mostly studied the motile stages of free-living dinoflagellates commonly found in freshwater, brackish, and marine plankton.

Dinoflagellates are also known to geologists as an important group of microfossils used extensively for biostratigraphy (e.g., in oil exploration). However, most fossil dinoflagellates are morphologically very different from the motile stages studied by biologists. Their natural affinities remained unknown until the early 1960s, when paleontologists discovered living equivalents and showed these to be nonmotile, benthic resting stages (cysts) of dinoflagellates. Since then, work mainly by micropaleontologists has shown that many living dinoflagellate life cycles routinely include such cysts.

Realization that paleontologists and biologists were independently studying two different stages in the life cycle of the same organisms has affected both fields of research. The past few years have seen a marked increase both in the paleontologists' interest in dinoflagellate biology and the biologists' interest in the previously neglected cysts. In particular, the study of living dinoflagellate cysts is emerging as an important field of research for both micropaleontology and phytoplankton biology. For the paleontologist, living cysts offer the exciting possibility for interpretations of the fossil record based on a better understanding of the organisms involved. For the biologist, living cysts present a challenging new "benthic view" of dinoflagellate ecology.

The study of living dinoflagellate cysts is a relatively new, rapidly growing field, where basic references are scattered throughout paleontological and phycological literature. As such, there is a particular need for "state of the art" reviews covering both aspects of the subject. The following summary of current knowledge concerning living dinoflagellate cysts is therefore presented in the hope that it may help paleontologists understand life processes ultimately producing the dinoflagellate fossil record and stimulate phycologists to further investigate the many remaining questions concerning the biological significance of cysts. This paper was first presented at a symposium "Survival Strategies in the Algae" organized by the Phycological Society of America and the Botanical Society of America in Vancouver, Canada, July 1980.

My aim is to present a general, comprehensive introduction to living dinoflagellate cysts rather than a review of all the literature. This summary of what the cysts are, how they are formed, and their possible significance regarding survival strategy is therefore a review of selected information from the literature, supplemented on occasion with unpublished observations from my own work. No attempt is made to review the ever-mushrooming literature on cyst terminology (though this is discussed in section 4.2) or the vast amount of stratigraphic literature, which is considered outside the scope of this work.

Previous reviews

Two reviews published a decade ago in the geological literature helped lay the foundation for subsequent living dinoflagellate cyst studies. The first (Evitt, 1970) stressed new developments in the field; the second (Wall, 1971) summarized much of the biological information known at that time. Information contained in these two works has also formed the basis for a review by Harland (1971), and sections on living cysts by Evitt (1969), Sarjeant (1974), Williams (1977, 1978), and Brasier (1980).

Quaternary dinoflagellate cyst studies have been reviewed by Wall (1970) and Reid and Harland (1977).

History of dinoflagellate cyst studies

The nonspecialist newly interested in living dinoflagellate cysts will have difficulty understanding the unusual interrelationship between biology and paleontology that lies behind the subject. As Evitt pointed out, in announcing a Penrose Conference on Modern and Fossil Dinoflagellates in 1978, the biologists and paleontologists involved in this field represent "two groups of scientists whose interests have been focused heretofore on different parts of the dinoflagellate life cycle and whose contacts and exchanges have been severely limited by contrasting professional backgrounds, affiliations, techniques, and goals." Understanding the history of dinoflagellate cyst studies is therefore fundamental to working in this subject.

The history of the study of fossil dinoflagellate cysts has been thoroughly reviewed by Sarjeant (1961, 1970, 1974) and Evitt (1970), whereas Reid (1978) provides some references to early observations of living cysts in plankton. Three main phases may be recognized within the roughly 150 years during which cysts have been studied: Phase 1, early discoveries; Phase 2, the accumulation of essentially separate bodies of information for fossil and living cysts; and Phase 3, the realization that biologists and paleontologists had largely been studying two different stages in the life cycle of dinoflagellates.

Phase 1 (1830–1920). Ehrenberg (1838) is credited with the first record of fossil cysts, seen in very thin flakes of Cretaceous flints. Although not realizing that these were dinoflagellate cysts, he noted two very different types. One showed characteristic shapes and sometimes even plate patterns, which Ehrenberg, who had also studied living plankton, had no difficulty identifying as dinoflagellates. He believed these to be fossilized thecae from planktonic motile stages. The other type, covered by well-developed spines (processes), showed no obvious dinoflagellate affinities; he thought these were the silicified zygospores of *Xanthidium* Ehrenberg, a genus of freshwater Desmids. Throughout the later nineteenth and early twentieth centuries, a few other scientists discovered similar spiny microfossils in flints and cherts. Suggestions for their affinities included statoblasts of freshwater Bryozoans and sponge spicules, though Mantell (1845) had correctly suggested that such fossils were probably organic rather than siliceous (see references in Sarjeant, 1961). When Reinsch (1905) stated that Ehrenberg's supposed "Xanthidia" were dinoflagellate cysts, the idea was apparently ignored.

Living dinoflagellate cysts were occasionally recorded in plankton studies of the late nineteenth and early twentieth centuries, though at that time their identities were often obscured. For example, the distinctive cyst of *Polykrikos schwartzii* Butschli (Fig. 25) was first recorded by Hensen (1887) as an Umrindete cyst, and its identity was further obscured by Lohmann (1910), who dissected cysts of this type but misinterpreted the repeated flagellar grooves of the dinoflagellates, thinking these were a metamerically segmented animal. In contrast, "spiny" dinoflagellate cysts were probably recorded together with "spiny" copepod eggs as *Trochischia* Kützing, or *Xanthidium*. Thus, for example, Cleve (1900) described one of these as *Xanthidium hystrix* Cleve, which he considered "probably a stage in the development of some dinoflagellate." However, Lohmann (1904) hatched copepod nauplii from what he considered *X. hystrix*. Unfortunately, many illustrations from this phase of investigation are so simplified that it is impossible to be sure which represent true dinoflagellate cysts.

Phase 2 (1920–1960). Paleontological research on cysts began to expand during the early 1930s. Wetzel created the genus *Hystrichosphaera* for the supposed "Xanthidia," whose real identity remained obscure, and they became popularly called hystrichospheres (hystrichosphaerids). Eisenack and the algologist Deflandre also figured prominently in this early phase of expansion, during which they and several others discovered and cataloged a wide range of hystrichospheres, largely from Mesozoic and Tertiary sediments. (See references in reviews by Sarjeant, 1970, and Evitt, 1970.) Two main factors prompted this expansion: First, the gradual development of techniques for chemically extracting and thereby concentrating the fossils from rock samples greatly increased possibilities for discovering and examining new forms. Second, the realization that hystrichospheres together with fossil plant spores and pollen grains obtained in such preparations were potential stratigraphic markers and paleoecological indicators gave rise to palynology (the study of these organic microfossils) as a "new tool" for the oil industry.

In sharp contrast to these paleontological developments, biological records of cysts continued to appear as occasional adjuncts to plankton surveys, which if at all interested in dinoflagellates, concentrated primarily on their motile stages. There were a few exceptional studies of living cysts, but these formed no cohesive body of literature comparable to that for the fossils. Huber and Nipkow (1922, 1923) produced a remarkably detailed study of encystment in the freshwater species *Ceratium hirundinella* (O.F. Müller) Bergh both in Lake Zurich and in laboratory cultures, whereas Braarud (1945) reported encystment in cultures of the marine species *Gonyaulax tamarensis* Lebour and *Protoceratium reticulatum* (Claparède & Lachmann) Bütschli, and Nordli (1951) documented encystment of a population of *Gonyaulax polyedra* Stein in the Oslofjord, Norway.

In addition to phycologists and palynologists working with geologically older sediments, a third group began taking interest in "the hystrichosphere problem" and dinoflagellate cysts. These were palynologists investigating the use of fossil pollen records to interpret paleoclimatic events within the Quaternary period. The first hystrichospheres ("Hystrix" as they were called by most Quarternary palynologists) discovered in pollen preparations were presumed to be reworked from older rocks. Pre-Quaternary palynological literature showed the abundance of such fossils in older rocks, and since no convincing link was established between these and living plankton, they were presumed to have become extinct sometime between Tertiary and Present times (e.g., Iversen, 1936). However, Erdtman (1949, 1950, 1954), reporting living hystrichospheres from sediment traps in a Swedish fjord, demonstrated clearly that they were extant. Wetzel identified *Hystrichosphaera furcata* Wetzel as one of Erdtman's living types, and Braarud (1945) identified others as dinoflagellate cysts. McKee et al. (1959) recovered rich assemblages of similar types from palynological preparations of Recent sediments in a Pacific atoll lagoon.

Phase 3 (1961–present). Undoubtedly, the turning point heralding the modern period of cyst studies was provided by Evitt (1961). He identified two fundamental morphological features of many hystrichospheres that were much more extensive and significant than had been recognized previously: the presence of a characteristic opening, which he called the archeopyle, and subtle "reflected" features from dinoflagellate motile stages (e.g., in some forms the number and position of processes reflected a typical dinoflagellate plate pattern). From this he concluded that most postpaleozoic hystrichospheres were probably dinoflagellate cysts that had formed inside generally nonfossilizable motile cells and the archeopyle represented an excystment aperture. As Evitt (1961, p. 403) pointed out, the idea that hystrichospheres could be dinoflagellate cysts was not new, since Braarud (1945, and in Erdtman, 1954) and Nordli (1951) had identified several before. However, in extending this to include many fossils, Evitt inspired a new way of looking at both fossil and living cysts.

Confirmation of Evitt's basic observations on fossils rapidly followed as several palynologists turned their attentions to living plankton. Evitt and Davidson (1964), examining preserved plankton tows from Oslofjord, found the three "living hystrichosphere" types previously reported by Braarud (1945), Nordli (1951), and Erdtmann (1954). They effectively linked the fossil record and living plankton by demonstrating diagnostic features such as the archeopyle in cysts of *G. polyedra* and *P. reticulatum*, and by identifying a type of *Hystrichosphaera* as the cyst formed within a *Gonyaulax digitalis* (Pouchet) Kofoid (from several remaining thecal plates attached to the cyst). At the same time, Wall and Dale were conducting the first experiments with living hystrichospheres recovered from bottom sediments. On incubation, these produced motile dinoflagellates, conclusively proving their role as resting cysts and for the first time revealing details of excystment (Wall, 1965; Wall & Dale, 1966).

Realization that the fossil record represented cysts whose living counterparts had hardly been studied has influenced dinoflagellate paleontology during the past decade or so. Fossil cysts have proved increasingly useful in pre-Quaternary stratigraphy and are now used extensively in oil exploration. This rapidly accumulating paleontological information has yielded biologically relevant ideas concerning, for example, dinoflagellate lineages (e.g., Wall & Dale, 1968a) and thecal plate homologies (e.g., Eaton, 1980). Progress in Quaternary cyst palynology has been comparatively slow from the promising early work of Rossignol (1961, 1962) until fairly recently, when notably Harland (1974, 1977), Morzadec-Kerfourn (1976), and Rossignol-Strick and Duzer (1979) began comparing Quaternary assemblages with known distributions of living cysts in attempted paleoecological interpretations. However, many Quaternary palynologists have continued to use the undefined term *Hystrix* for some cysts while ignoring most others.

Both palynologists and biologists have contributed to studies of living cysts during this period (see references in later sections of this chapter). Palynologists have mainly investigated the morphological range of living cysts, which dinoflagellates produce them, and their ecological distribution. Biologists have mainly studied the role of cysts in sexual cycles and in bloom phenomena, particularly toxic forms producing red tides.

4.2 Cyst terminology and morphology

Before proceeding further, it is necessary to address at least some of the main problems of terminology. Essentially these fall into two main categories. First, we must define what is meant by the term *resting cyst* as used here and particularly show how this relates to various similar terms used in biology and paleontology. Second, we need to consider descriptive terminology applied to cyst morphology.

Use of the term cyst in dinoflagellate studies

Table 1 summarizes the main use of cyst and related terms in biological and paleontological literature concerning dinoflagellates. In paleontology, there is generally good agreement as to what is meant by the term *dinoflagellate cyst*. It is now widely accepted that all known fossil dinoflagellates are the remains of nonmotile resting stages (cysts), the main diagnostic features of which are shown in Fig. 1 and summarized in Table 1. Biologically, the fossils probably functioned as hypnozygotes (see discussion in section 4.3 under Cysts and sexual cycles). However, since this cannot be verified in the fossil record, the term *hypnozygote* is inappropriate in paleontology. Similarly, there is no need for the term *resting cyst* in paleontology, and *dinoflagellate cyst* remains an adequate, generally accepted term for the fossils. But to avoid unwieldy repetition of "dinoflagellate cyst" in publications, some palynologists use the term *dinocyst* (from Lentin & Williams, 1976), whereas others simply explain that *cyst* alone is used in the interest of brevity.

Thus, in paleontology the terms *dinoflagellate cyst*, *dinocyst*, and *fossil dinoflagellate* are usually synonyms, and there is no basis for recognizing

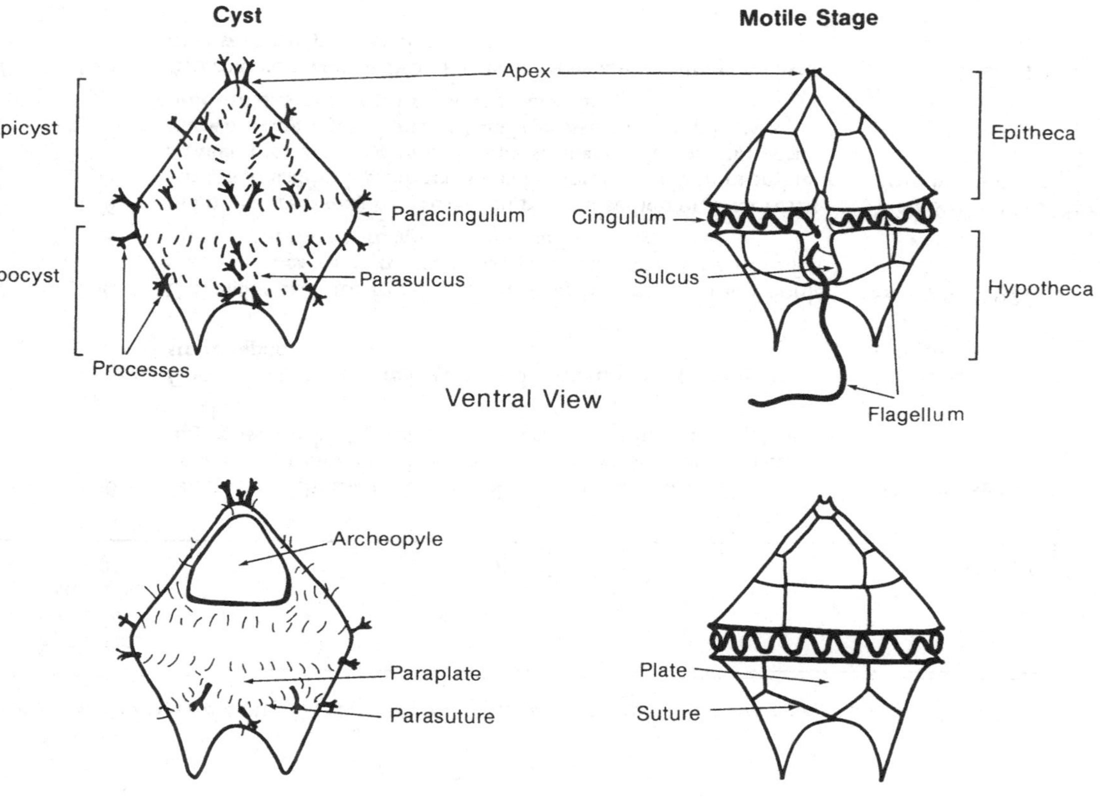

Fig. 1. Comparison of the main features of resting cyst and thecate motile stage in dinoflagellates.

Table 1. *Summary of the main uses of dinoflagellate cyst and related terms in biological (B) and paleontological (P) literature*

Cyst term	Usage P	Usage B	Synonyms or overlapping terms	Comments	References
1. Dinoflagellate cyst	X	/	1–6	Most generally used term in paleontology for all fossil dinoflagellates – often abbreviated to "cyst" in publication; usually too imprecise for biological use – necessary to differentiate Nos. 3–10.	Widely used
2. Dinocyst	X		1–6	Convenient abbreviated term used particularly by some biostratigraphers.	Lentin & Williams, (1976);
3. Fossil dinoflagellate	X	/	1–6	Originally used for fossils presumed to be fossil thecae (motile stages) – these now regarded as cysts with reflected motile features; term therefore applicable to all fossil cysts	Sarjeant, (1974)
4. Resting cyst	/	X	1–6	Main features: may be present in bottom sediment, wall resistant (chemical/biological decay), cell contents include prominent storage products (e.g., starch, lipids), mandatory resting period; known living equivalents of fossils shown to be type cyst – other fossils presumed to be type cyst (here)	Wall & Dale (1968a); Fritsch (1956)
5. Resting spore	X	/	1–6	Alternative term for No. 4 – used especially in literature associated with early incubation experiments	Wall & Dale (1970)

6. Hypnozygote		X	1–6	Originally used for proved resting zygotes in sexual cycles of several nonfossilizable freshwater forms; similarity between these and sexual cycles of fossilizable marine forms noted – fossils therefore presumed to be equivalent to hypnozygotes (here)	von Stosch (1972, 1973); Dale (1976)
7. Acritarch	X		1–6 (at least one)	Lacking obvious dinoflagellate (or otherwise identifiable) affinities; probably heterogenous group, but incubation experiment proved one "living acritarch" to be a dinoflagellate cyst	Evitt (1963); Dale (1977a)
8. Temporary cyst		X	9, ?10	Nonmotile stage formed by some dinoflagellates in cultures subjected to adverse conditions (e. g., temperature); so far not recorded from sediments; differs from No. 4 by generally not so resistant wall, less storage products in contents, no observed mandatory resting period (name based on these differences)	Dale (1977b)
9. Pellicle cyst		X	9, ?10	Alternative term for No. 8 – based on presumed cyst wall equivalence to pellicle layer of motile stage	Anderson & Wall (1978)
10. Ecdysal cyst		X	?8, 9	Nonmotile cell remaining after ecdysis (i.e., shedding of thecae and flagella by some motile stages, particularly in response to adverse conditions); relationship between this and other cyst types never clarified.	Taylor (1980)
11. Dinospore		X	?	Temporary motile stage in asexual life cycle of certain dinoflagellates (e.g., *Dissodinium pseudocalani* [Gönnert] Drebes)	Elbrächter & Drebes (1978)

different types of cysts as required in biological studies. More "biological" terms, such as *resting cyst* or *resting spore*, are sometimes encountered in paleontological literature (especially where living equivalents of fossils are known), but in this context these, too, may be considered synonymous with the other fossil cysts. Organic-walled microfossils not identified as dinoflagellate cysts and lacking other recognized affinities are termed *acritarchs*, an artificial *incertae sedis* group erected for taxonomic convenience (Evitt, 1963). The acritarchs were thought to include remains of organisms of different biological affinities, possibly including some dinoflagellate cysts lacking the usual definitive morphological features (Lister, 1970). This was confirmed recently when the living equivalent of one fossil acritarch was shown to be the resting cyst of the dinoflagellate *Peridinium faeroense* Paulsen (Dale, 1977a).

From a biological viewpoint, using the term *cyst* for dinoflagellates invites confusion, since this has been widely used to describe a variety of nonmotile stages in other organisms, notably Protozoans. The most frequently used alternative terms in biological literature are *dinoflagellate spore* or *resting spore*. In contrast to fossil cysts, the many different nonmotile dinoflagellate stages so far encountered in biological studies are not adequately covered by a single term.

A detailed review of all reported nonmotile stages is beyond the scope and purpose of this chapter. From the often fragmentary evidence available, two main types are recognized here, termed *resting cysts* and *temporary cysts*, based on their observed or presumed mode of functioning (Dale, 1977b). Resting cysts are the better documented and form the main theme for this chapter. Their basic features are described later in this section and section 4.3 and are summarized in Table 1. Both morphology and physiology attest to their resting function, probably as hypnozygotes in a sexual cycle.

The term *temporary cyst* is used informally here as a general term for nonmotile stages other than resting cysts. As such, it probably encompasses several different stages in dinoflagellate life cycles, including the following:

1. *Stages to withstand temporary adverse conditions* (e.g., of temperature). The most well-documented example is that of *G. tamarensis* or *Gonyaulax excavata* (Braarud) Balech (Fig. 2), cultures of which have been shown to produce temporary cysts when cooled below 5°C (Prakash, 1963; Dale, 1977b; Anderson & Wall, 1978). Temporary cysts differ from resting cysts in these species by their lacking obvious storage products and a mandatory resting period. (Temporary cysts regenerate motile cells on return of favorable conditions.) These have been called *pellicle cysts* by Anderson and Wall (1978), from the presumed correspondence between the cyst wall and the pellicle layer of the motile stage.

2. *Stages resulting from ecdysis* (a characteristic shedding of theca by some motile dinoflagellates). This may also be in response to unfavorable conditions, in which case these stages may be equivalent to No. 1. However, Taylor (1980) has called these ecdysal cysts.

3. *Stages in vegetative (asexual) reproduction*. In its simplest form, this may be a temporary nonmotile stage allowing a cell to divide having first cast off a theca (e.g., *Pyrophacus horologium* Stein, Fig. 3).

Apart from lacking the distinctive characteristics of resting cysts, temporary cysts seem to be more transitory in nature where they are sometimes seen in plankton associated with motile stages but not in bottom sediments, where the resting cysts accumulate.

The differences in cyst types and terminology are far from just an academic problem. Particularly in the biological literature, the reader needs to differentiate between cyst types that may be confused with each other. For example, in red-tide work by Prakash (1963), Bourne (1965), and Prakash et al. (1971), it is important to realize that the cysts referred to are temporary cysts (known so far only from cultures) rather than resting cysts, which are now well documented from nature but were unknown to these authors (Dale et al., 1978, p. 1224). A further example is one of the arguments that Reid (1974, p. 584) used for maintaining separate classification systems for cysts and motile stages: "The creation of holomorphic taxa, as advocated by Wall and Dale (1968a) would seem to be unrealistic in the light of present information on the complicated life cycles of dinoflagellates. Up to the present day only two detailed studies on dinoflagellate life cycles have been made, both on the same species and both giving contrasting results (Buchanan, 1968; Wall & Dale, 1969) Fig. 2." In this case, it is important to realize that the "contrasting results" are largely due to comparing temporary cysts (Buchanan) with resting cysts (Wall & Dale). Obviously, it is important to differentiate these in defining holomorphic taxa.

For the remainder of this chapter, the term *cyst* will be used instead of *dinoflagellate resting cyst*, for the sake of brevity.

Cyst morphology

Cysts and motile stages occurring in the life cycle of the same dinoflagellate may be considered as two different morphological expressions of one set of genes. To a large extent, these morphological differences reflect the obviously different functions of a motile stage and a nonmotile resting stage, but a surprisingly wide range of morphological "overlap" exists between the two. In a broad sense, this allows us to recognize: (1) "typical" cyst features (e.g., the archeopyle – an excystment aperture), (2) "typical" motile stage features (e.g., cingulum and sulcus –

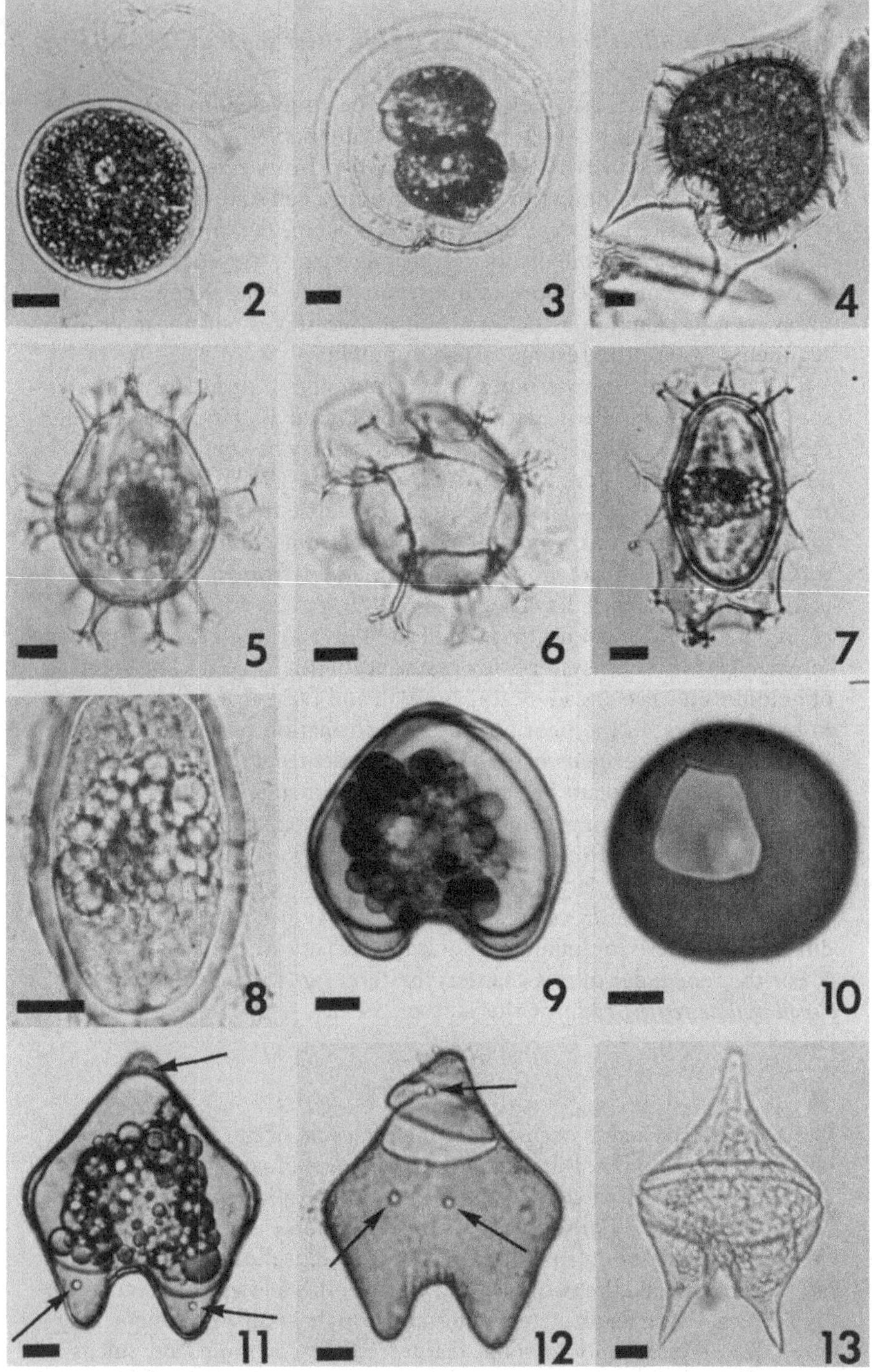
2
3
4
5
6
7
8
9
10
11
12
13

Figs. 2–13. Scale bar = 10 μm. **Fig. 2.** Temporary cyst in culture of *Gonyaulax excavata* from the Oslofjord cooled overnight from 15° to 5°C. **Fig. 3.** Temporary cyst of *Pyrophacus horologium* with dividing protoplast, from Oslofjord plankton August 29, 1978. **Fig. 4.** Newly formed resting cyst of *Protoperidinium claudicans* with theca still attached, from Oslofjord plankton September 12, 1978. (Photo kindly supplied by Karl Tangen.) **Fig. 5.** Live resting cyst of *Gonyaluax digitalis* from bottom sediments in the inner Oslofjord March 3, 1976. **Fig. 6.** Empty resting cyst of *Gonyaulux scrippsae* (incubation experiment DC 139) from bottom sediment in the Bay of Vigo, Spain, September 30, 1977. **Fig. 7.** Live *Gonyaulax spinifera* group resting cyst (paleontological name: *Spiniferites elongatus*) from bottom sediments in the inner Oslofjord March 11, 1975. **Fig. 8.** Details of live cell contents (predominantly starch grains) in *Spiniferites elongatus* from bottom sediments in Somes Sound, N.E. USA, November 23, 1975. **Fig. 9.** Live *Protoperidinium oblongum* resting cyst from bottom sediments in the inner Oslofjord March 4, 1976. Cell contents predominantly lipid globules. **Fig. 10.** Empty resting cyst of *Protoperidinium conicoides* from bottom sediments near Drøbak, Oslofjord, June 27, 1974. **Fig. 11.** Live *Protoperidinium oblongum* resting cyst (incubation experiment DC 134) from bottom sediment in the Bay of Vigo, Spain, September 30, 1977. **Fig. 12.** Empty cyst (as Fig. 11) after incubation. Note granules (arrowed) of unknown function. **Fig. 13.** Thecate motile stage of *Protoperidinium oblongum* from incubation of cyst in Figs. 11 and 12.

grooves housing the flagella), and (3) morphologically "overlapping" features (e.g., paracingulum and parasulcus – features on a cyst resembling flagellar grooves of the motile stage but not functioning as such since the cyst has no flagella).

Understanding dinoflagellate motile stage features is thus a necessary step toward understanding cyst morphology. Fig. 1 therefore compares some of the main features of cysts and motile stages. The overlapping features between cysts and motile stages pose a particularly difficult conceptual problem. Almost all such overlap involves motile stage features reflected to some extent by cyst morphology, although in at least one genus, *Cladopyxis* Stein, the motile dinoflagellate looks more like a cyst. Extreme examples, such as the fossil cysts *Palaeoperidinium pyrophorum* (Ehrenberg) Sarjeant (Gocht & Netzel, 1976), the calcareous cysts *Calciodinellum* (Figs. 29–31), or the siliceous *Peridinites* (Fig. 26), reflect motile stage features to such a degree that until recently they were believed to be fossilized thecae (e.g., Ehrenberg's original fossil dinoflagellates). Many others, while predominately showing clear cyst features, nevertheless include features reflecting flagellar grooves, flagellar pores, and thecal plates.

As yet, we know little about how or why such motile stage features occur on cysts. Observations on encystment in living dinoflagellates have certainly not supported earlier theories of direct physical control (mechanical imprinting) of cyst morphology by the parental thecal wall (see discussion in section 4.3 under Encystment), and the term *reflect* as used here is not meant to imply this. Cyst morphology is distinctive, constant within a species, and generally less conservative than that of corresponding motile stages. This strongly suggests direct genetic control of cyst morphology comparable to that for motile stages. If so, morphological overlap possibly reflects degrees of genetic separation of factors responsible for cyst building versus motile stage building within the total gene set for a given species. The term *reflect* as used here is an attempt to express this admittedly poorly understood morphological relationship, for example, that a paracingulum on a cyst somehow genetically reflects the fact that the corresponding motile stage has a cingulum.

It would take a far larger book than this to include all the (probably hundreds of) descriptive terms used for cysts. Fortunately, many of these are brought together in a useful glossary by Williams, et al. (1978). Further extensive modifications in terminology, especially dealing with morphological overlap between cyst and motile stage, are recommended by Evitt et al. (1977). Within palynology there is a major disagreement on the need for separate terms for reflected motile stage features on cysts. Evitt et al. (1977) express the need for separate terms using largely the prefix "para" for cyst features (e.g., paracingulum, Fig. 1), while Norris (1978, p. 303) thinks this "needlessly complicating with technical verbiage a basically simple relationship," since "none now seriously believe that thecae are represented in the fossil record [with rare exceptions discussed later] and no confusion or false implications are likely using the term plate for both cysts and thecae." Norris's criticism reflects the frustrations of some paleopalynologists working exclusively with fossils, but some of the "para" terminology of Evitt et al. is certainly useful as used here in discussions of both motile stages and cysts.

The main morphological features used in identifying or adequately describing cysts are briefly described.

General body shape. Body shape is one of the most conservative cyst features. As seen in the figures presented here, overall body shape of cysts may vary considerably from species to species, but it remains generally constant within a species and often falls into one of two main morphological categories. These are (1) cyst body shape obviously related to that of the motile stage (e.g., cysts of the freshwater species *C. hirundinellum* with characteristic ceratioid horns), in some cases body shape is diagnostic, for example, allowing recognition of living and fossil peridinioid cysts (e.g., Figs. 14–16), and fossil gymnodinioid cysts (Evitt, 1967a); (2) basic spherical cyst body shape, this seems to be a particularly conservative shape, including examples from many different groups of dinoflagellates (e.g., within the major genera *Protoperidinium* Bergh, Figs. 10, 19, 21; *Gonyaulax* Diesing, Figs. 36, 37; and the unarmored dinoflagellate genera *Gymnodinium* Stein and *Gyrodinium* Kofoid et Swezy, Fig. 23).

Cysts representing both the categories mentioned and occasionally others may occur within one closely related group of dinoflagellates. This is one of the complicating factors between the paleontological, cyst-based classification system for dinoflagellates and the biological, motile stage-based system (discussed in section 4.4 under Living cysts and classification). For example, within the *Gonyaulax spinifera* group (i.e., dinoflagellates with thecae closely resembling that in Fig. 45), cyst body shape is often very similar to that of the motile stage (Fig. 5), but several basic spherical types are known (Figs. 36, 37) and at least one example is known in which the cyst is distinctly elongate (Figs. 7, 8), but the motile stage is not.

Overall body shape and particularly the position of horns, paracingulum, and parasulcus are useful features for "orienting" a cyst, such as recognizing apex, antapex, and dorsal and ventral surfaces (Fig. 1).

Cyst wall structure. Cysts are usually enclosed by one, two, or three moderately thick to very thick walls. These are called *endophragm, mesophragm,* and *periphragm,* and the spaces enclosed are called *endocoel, mesocoel,* and *pericoel* (see Evitt et al., 1977, Tables 2, 3, Fig. 2). In most

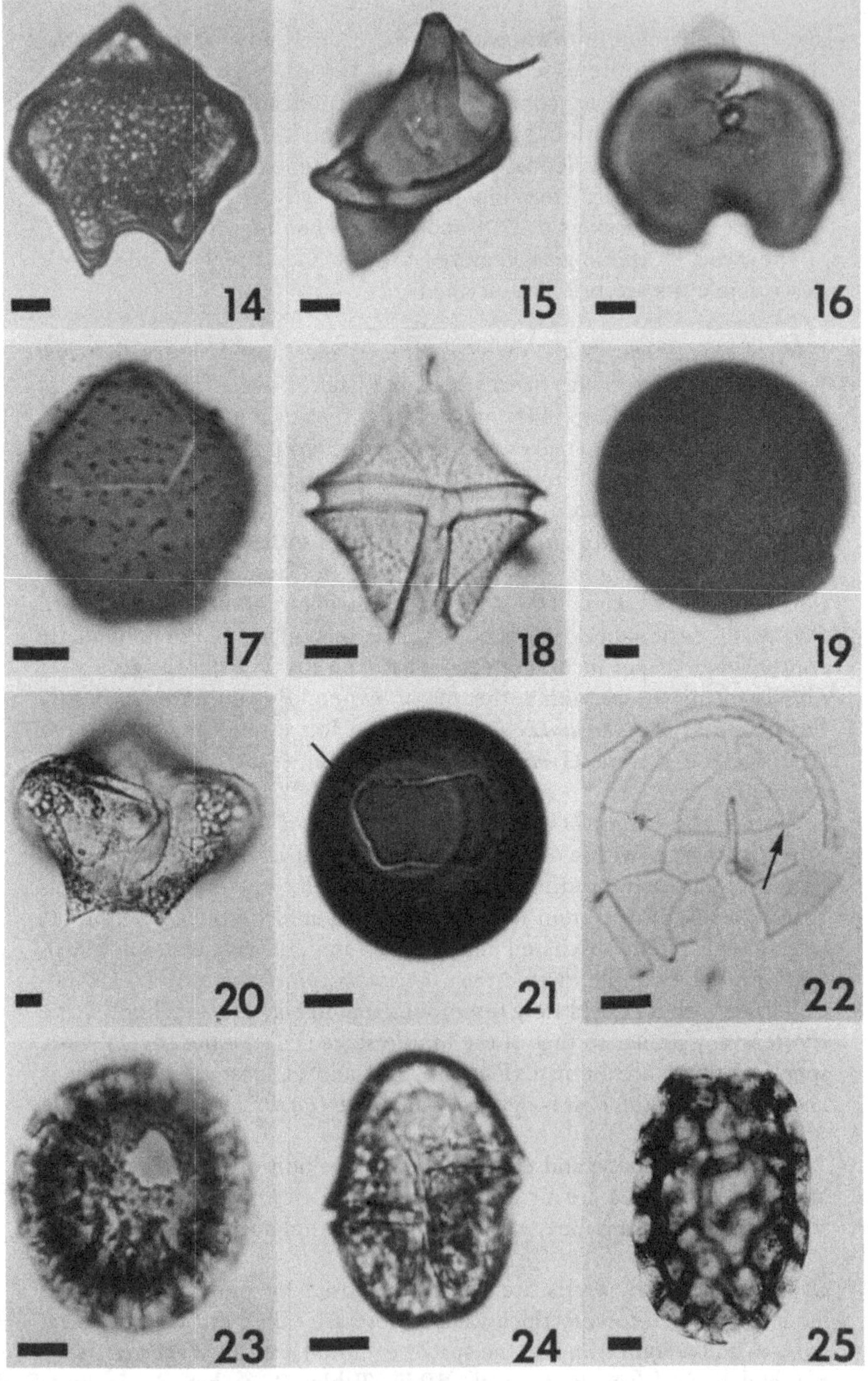
14
15
16
17
18
19
20
21
22
23
24
25

Figs. 14-25. Scale bar = 10 μm. **Fig. 14.** Live ***Protoperidinium leonis*** resting cyst (incubation experiment DC 150) from bottom sediment in the inner Oslofjord March 20, 1978. **Fig. 15.** Empty cyst (as Fig. 14) after incubation; lateral view. **Fig. 16.** Same cyst (as Fig. 15) seen in polar view. **Fig. 17.** Empty cyst (paleontologic name: ***Peridinium ponticum***) in incubation experiment DC 45; from bottom sediments in the inner Oslofjord March 4, 1976. **Fig. 18.** Theca of ***Protoperidinium*** cf. ***divergens*** in incubation experiment DC 70; from identical cyst to Fig. 17; from bottom sediments in the inner Oslofjord March 4, 1976. **Fig. 19.** Empty cyst (paleontologic name: ***Brigantedinium majusculum***) in incubation experiment DC 171; from bottom sediment in the Bay of Vigo, Spain, September 30, 1977. **Fig. 20.** Live thecate ***Protoperidinium*** sp. from cyst in Fig. 19 (incubation experiment DC 171). **Fig. 21.** Empty resting cyst of ***Protoperidinium denticulatum*** (incubation experiment DC 188) from bottom sediment off Monhegan Island, N.E. USA, January 1978. **Fig. 22.** Epitheca from one of a pair of motile ***Protoperidinium denticulatum*** from cyst in Fig. 21 (incubation experiment DC 188). Arrows show a side of thecal plate 1a and its equivalent on the corresponding operculum of the cyst. **Fig. 23.** Empty resting cyst of ***Gyrodinium resplendens*** (incubation experiment DC 202) from bottom sediment in Chesapeake Bay, E. USA, November 1978. **Fig. 24.** Live ***Gyrodinium resplendens*** from identical cysts to Fig. 23 from the same sediment. **Fig. 25.** Empty resting cyst of ***Polykrikos schwartzii***; Recent sediment from the Trondheimsfjord, Norway.

species, cyst walls are composed of organic matter, but mineralized cysts are well known from the fossil record and from a few living representatives.

Organic cyst walls may be variously resistant to both natural decay and laboratory acid treatments, depending on the species. Many are extremely resistant (the "acid-resistant" cysts studied by palynologists) and are thought to fossilize because of sporopolleninlike material in the wall (Brooks & Shaw, 1973). These authors have shown that sporopollenins are a unique class of biopolymers formed by an oxidative polymerization of carotenoids and/or carotenoid esters. They are found in resistant pollen grain and spore walls in higher plants and in spores of some algae and fungi. Cyst walls in many living dinoflagellates, however, are less resistant and would probably not fossilize. Among these are many cysts of heterotrophic *Protoperidinium* species apparently lacking photosynthetic pigments, raising the question of whether these species differ biochemically by not synthesizing carotenoids and are thus unable to produce sporopollenins (as in some fungi reported by Brooks & Shaw, 1973). The less resistant *Protoperidinium* cysts are usually a distinct brown to dark brown color, whereas sporopolleninlike cysts are mostly colorless, further suggesting possible biochemical differences. Even within one cyst, various parts may show different degrees of resistance. For example, Wall and Dale (in Wall et al., 1973) described a peridinioid cyst covered with small spines that were easily removed by acid treatments, which otherwise left the cyst wall intact (Fig. 17), whereas Manum (1978) described a Tertiary fossil representing the opposite situation, in which the cyst wall had not fossilized but elements representing paratabulation had fossilized (Fig. 38). As yet little work has been done on the ultrastructure of cyst walls, probably due to difficulties with fixing and preparation, but Bibby and Dodge (1972) described wall ultrastructure of *Woloszynskia tylota* (Mapletoft et al.) Bibby et Dodge cysts, Jux (1976) has studied thin sections of several Holocene cysts that have living representatives, and Dürr (1979) has studied ultrastructure of living *Peridinium cinctum* (O. F. Müller) Ehrenberg cysts by means of freeze-etching techniques.

Mineralized cysts are not so well known as organic-walled cysts, largely because they are destroyed by hydrochloric acid (HCl) and hydrofluoric acid (HF) used routinely in palynological preparation methods. Calcareous cysts were first described from Mesozoic and Tertiary rocks by Deflandre (1947, 1948), who understandably thought these were fossilized thecae (see Figs. 29, 31). Wall and Dale (1968b) described fossil calcareous cysts from Quaternary sediments and living calcareous cysts from bottom sediments in the ocean. Wall et al. (1970), using electron diffraction techniques, showed that at least one of these was composed of calcite. Incubation experiments with several types sug-

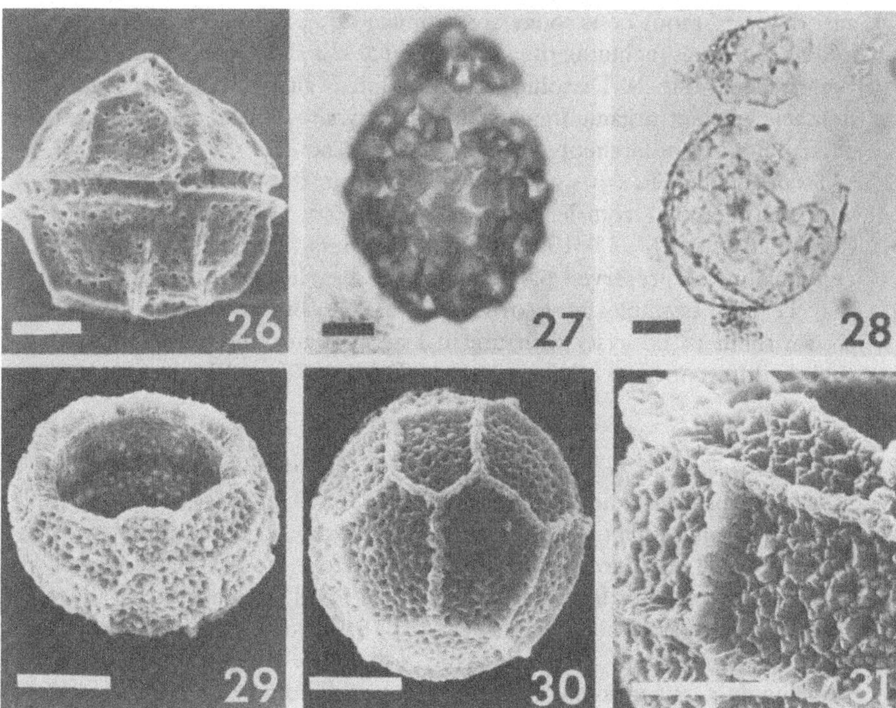

Figs. 26–31. Scale bar on Fig. 31 = 5 μm, others = 10 μm. Figs. 26 and 29–31 are SEM photographs. **Fig. 26.** Siliceous fossil dinoflagellate cyst *Peridinites globosus* from Eocene sediment in the Norwegian Sea. **Fig. 27.** Empty calcareous dinoflagellate cyst (so far unidentified) from incubation experiment DC 117, bottom sediment near Tromsø, N. Norway, December 5, 1977. **Fig. 28.** Organic membrane with traces of tabulation remaining after cyst in Fig. 27 digested in dilute HCl. Stained by trypan blue. **Fig. 29.** Calcareous dinoflagellate cyst (*Calciodinellum operosum*) from sediment trap at 988 m at PARFLUX Station E, 12°N, 54°W Tropical North Atlantic, November 1977 to February 1978. Ventral view. **Fig. 30.** *Calciodinellum operosum* in antapical view from same sample as Fig. 29. **Fig. 31.** *Calciodinellum operosum* showing details of crystaline calcareous wall; from same sample as Figs. 29 and 30.

gest that calcareous cysts today are produced by a closely related group of dinoflagellates including the genera *Scrippsiella* Balech and *Ensiculifera* Balech (see Table 2). Dissolution of the outer calcareous wall reveals a thin acidresistant organic inner wall that may show clearer evidence of paratabulation and archeopyle, as in Figs. 27 and 28.

To date, siliceous cysts are known only from the fossil record. They were first recorded from Tertiary rocks by Lefévre (1933) and Deflandre (1933); Eisenack (1935, 1936) described others from the Jurassic. Exceptionally well-preserved paratabulation led earlier workers to believe that Tertiary examples were fossilized thecae, but reexamination has shown them to be cysts with unusual archeopyles and paratabulation slightly different from calcareous cysts (Dale, 1978, and in preparation, Fig. 26).

Cyst wall surface features. Cyst walls may be smooth and unornamented, as in many protoperidinioid cysts (e.g., Fig. 9–12), or carry a wide morphological range of projections and ornamentation. These outgrowths from the wall often are distinctive and therefore are useful taxonomic criteria. The many different types found on fossil cysts are amply illustrated in the glossary of Williams et al. (1978). Two in particular are well represented in living cysts: ridges and processes. Ridges often outline paracingulum or parasulcus (e.g., in protoperidinium cysts, Figs. 14–16) or the margins of paraplates, when they are termed *parasutural ridges* (e.g., in gonyaulacoid cysts, Figs. 6, 33, 36). Processes may define paraplates by their number or position (e.g., Figs. 5–7, 33–35) or bear no apparent relationship to paratabulation (e.g., Figs. 4, 43, 44). The type of process is usually distinctive (i.e., whether solid or hollow, flat-tipped, flared, or pointed), though in some species the length of processes may vary considerably within a population (e.g., cysts of *Gonyaulax grindleyi* Reinsch). The cyst of *G. polyedra* [called *Lingulodinium machaerophorum* (Deflandre et Cookson) Wall in paleontology] is a notable exception, often developing characteristic clavate processes rather than the usual more pointed ones in response to very low salinity environments (Wall et al., 1973).

Sarjeant and Downie (1966) introduced terms for informally classifying cysts into three main morphological groups. One group, cavate cysts, was defined as having two wall layers with a sizable space between, the pericoel. Although important in the marine fossil record, this group is represented today by only a few freshwater forms with comparatively small pericoels. The other two groups were defined on the basis of a theory of contractional growth of cysts within thecae. The authors reasoned that cysts with proportionally longer processes formed by a greater degree of contraction from the thecal wall than did cysts with little or no such projections (Sarjeant, 1965). Cysts presumed to have formed close to the thecal wall and thought to resemble the theca more

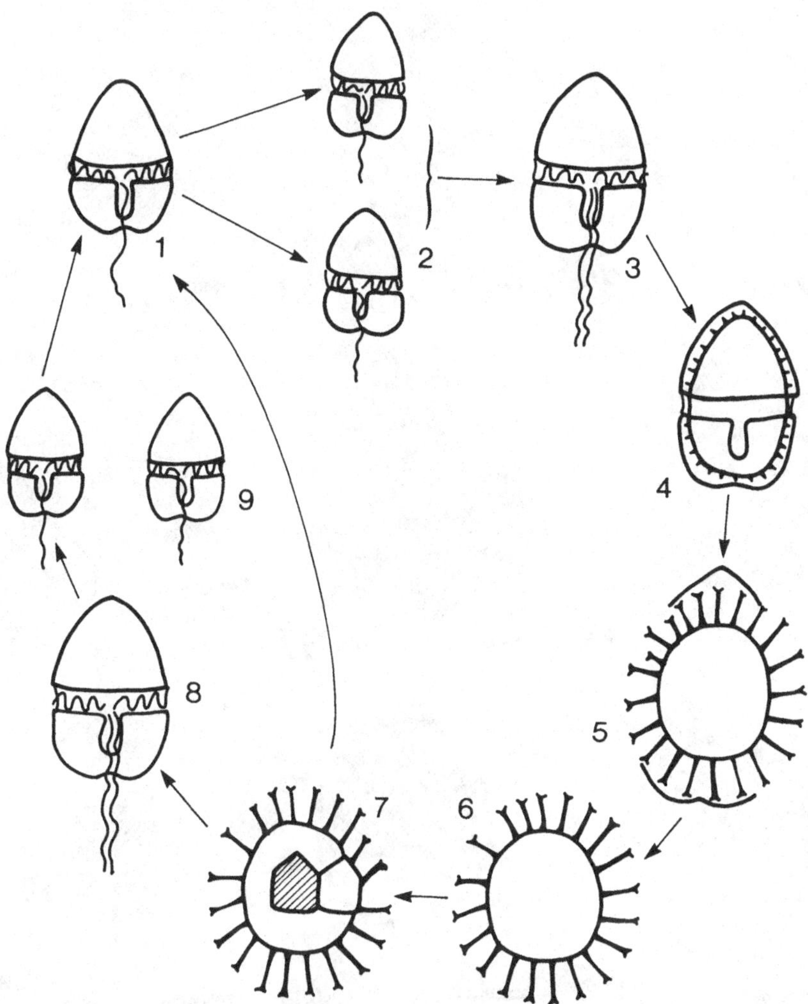

Fig. 32. Schematic diagram of the basic dinoflagellate sexual cycle producing cysts as hypnozygotes (based largely on work by von Stosch). Motile planktonic vegetative stage (1), producing gametes (2) pairs of which fuse to give planozygote (3). This eventually loses motility and cyst formation begins (4) and may proceed (5) by expansion rather than contraction of cyst (see text), producing resting cyst (hypnozygote) (6). Excystment (7) may produce a planozygote (8), comparable with (3), which by reduction division (9) reestablishes planktonic vegetative stage (1), or in other species the reduction division is completed during encystment, allowing direct reestablishment of stage (1).

closely were termed *proximate cysts*; those presumed to have formed well away from the thecal wall were termed *chorate cysts*. Observations on encystment in living dinoflagellates have not supported this theory (see discussion in 4.3 under Encystment), and since many cysts cannot readily be subdivided into two such groups based on relative extent of cyst wall outgrowths, the terms *proximate* and *chorate* are of little value.

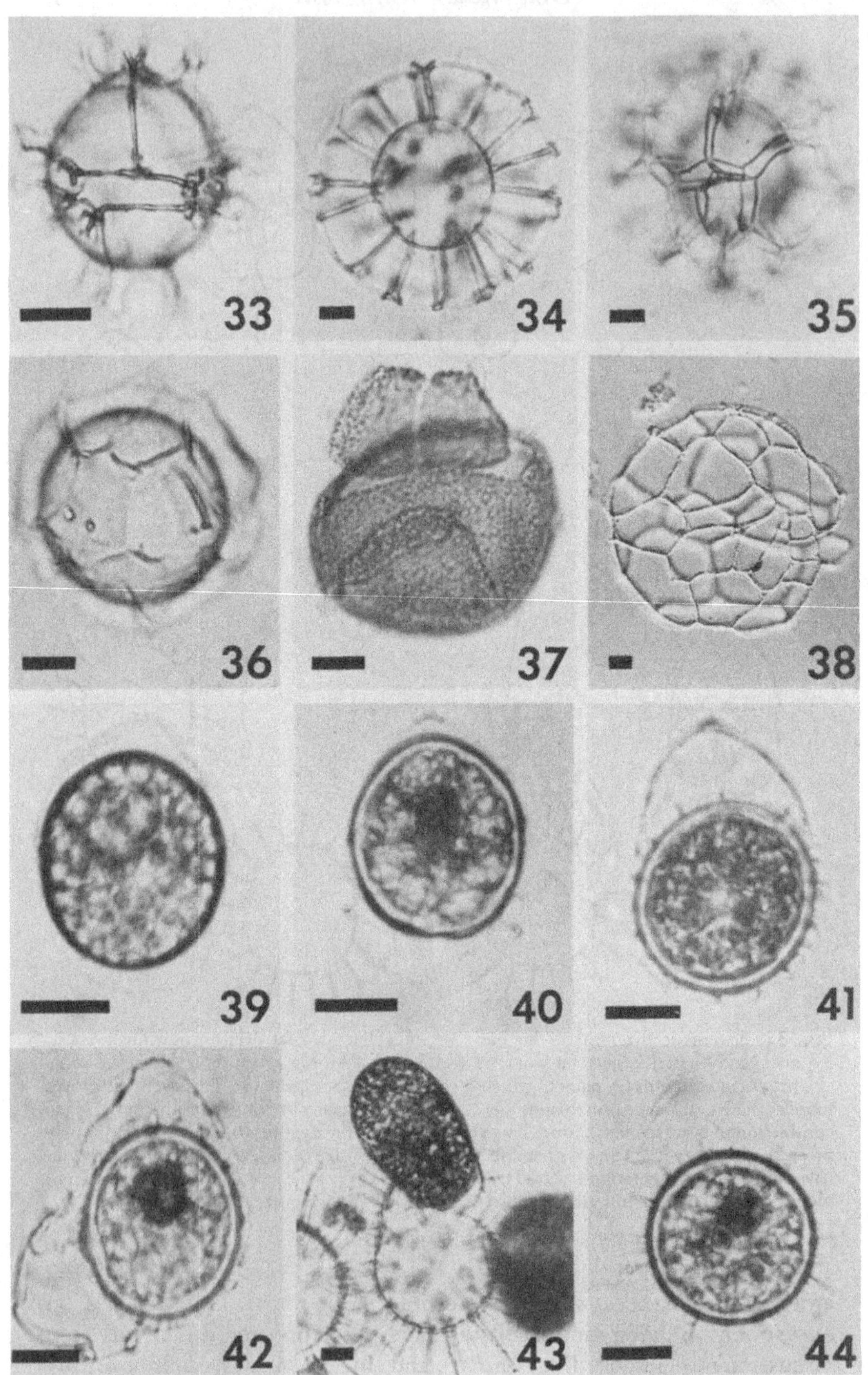
33
34
35
36
37
38
39
40
41
42
43
44

Figs. 33–44. Scale bar = 10 μm. **Fig. 33.** Empty resting cyst of *Gonyaulax scrippsae* (incubation experiment DC 139) from bottom sediment in the Bay of Vigo, Spain, September 30, 1977. **Figs. 34–35.** *Gonyaulax spinifera* group resting cyst (paleontological name: *Nematosphaeropsis labyrinthus*) with process tips joined by thin strands (trabeculae). From plankton tow near the Shetland Isles, March 30, 1966. **Fig. 36.** Empty *Gonyaulax spinifera* group cyst (*Spiniferites* sp.) with archeopyle equivalent to two paraplates (2p), from incubation experiment DC 130. From bottom sediment near Tromsø, N. Norway, December 5, 1977. **Fig. 37.** Empty *Gonyaulax spinifera* group cyst (*Bitectatodinium tepikiense*) showing both opercular pieces and the 2p archeopyle. Recent sediment from the Trodheimsfjord, Norway. **Fig. 38.** Fossil dinoflagellate cyst *Evittosphaerula paratabulata* from Miocene sediments in the Norwegian Sea. (SEM photograph kindly supplied by S. B. Manum.) **Figs. 39–42, 44.** Stages in cyst formation recorded from an encysting culture of *Peridinium faeroense* (DC 36) started from cysts in bottom sediments of the inner Oslofjord, March 4, 1976. Photographs are of different specimens. **Fig. 39.** Planozygote shortly after loss of motility. **Fig. 40.** Early stage of cyst formation with prominent pigmented body in cell contents. **Fig. 41.** Early stage of process formation in hyaline gelatinous layer. Note immature cyst body width approximately equal to original planozygote theca width seen on attached epitheca. **Fig. 42.** Second example of early process formation showing relative width of cyst body and attached planozygote theca. **Fig. 43.** Excysting *Gonyaulax polyedra* in culture experiment with cysts in bottom sediment from the Bay of Vigo, Spain, September 30, 1977. **Fig. 44.** More mature cyst with attached epitheca. Cyst body width remains equal to thecal width, whereas processes have grown out to about 50% more. Compare with Fig. 41.

Table 2. *Summary of known cyst/motile stage correlations*

Marine species with organic-walled cysts	References	Paleontological name for cyst
Diplopeltopsis minor	Wall & Dale (1968a)	*Dubridinium caperatum*
Diplopsalis lenticula	Wall & Dale (1968a)	
Diplosalopsis orbicularis	Wall & Dale (1968a)	*Dubridinium cavatum*
Gonyaulax digitalis	Wall & Dale (1968a)	*Spiniferites bentori*
G. excavata	Dale (1977b)	
G. grindleyi	Wall & Dale (1968a)	*Operculodinium centrocarpum*
? *G. grindleyi*	Wall & Dale (1970)	*O. israelianum*
? *G. grindleyi*	Wall & Dale (1968a)	*O. psilatum*
G. monilata	Walker & Steidinger (1979)	
G. polyedra	Wall & Dale (1968a)	*Lingulodinium machaerophorum*
G. scrippsae	Wall & Dale (1968a)	*Spiniferites bulloideus*
G. spinifera	Wall & Dale (1968a)	*S. mirabilis*
G. spinifera	Wall & Dale (1968a)	*Nematosphaeropsis labyrinthus*
G. spinifera	Wall & Dale (1968a)	*Tectatodinium pellitum*
G. spinifera group (undifferentiated)	Dale (1976)	*Bitectatodinium tepikiense*
G. spinifera group (undiff.)	Dale (1976)	*Planinosphaeridium membranaceum*
G. spinifera group (undiff.)	Dale (1976)	*Spiniferites elongatus*
G. spinifera group (undiff.)	Dale (1976)	*S. membranaceus*
G. spinifera group (undiff.)	Wall & Dale (1970)	*S. ramosus*
G. spinifera group (undiff.)	Wall & Dale (1968a)	*S. scabratus*
G. spinifera group (undiff.)	Dale (unpublished) (Fig. 36)	
G. tamarensis	Anderson & Wall (1978)	
Gymnodinium species	Wall & Dale (1968a)	
Gyrodinium resplendens	Dale (unpublished) (Figs. 23,24)	
? *Gyrodinium* sp.	Wall & Dale (1968a, Plate 4, Fig. 29)	
Peridinium faeroense	Dale (1977a)	
Protoperidinium avellana	Wall & Dale (1968a)	*Brigantedinium cariacoense*
P. claudicans	Wall & Dale (1968a)	*Votadinium spinosum*
P. compressum	Wall & Dale (1968a)	*Stelladinium stellatum*
P. conicoides	Wall & Dale (1968a)	*Brigantedinium simplex*
P. conicum	Wall & Dale (1968a)	*Multispinula quanta*
P. denticulatum	Wall & Dale (1968a)	
P. excentricum	Wall & Dale (1968a)	
P. latisimum	Wall & Dale (1968a)	
P. leonis	Wall & Dale (1968a)	*Trinovantedinium sabrinum*
P. minutum	Wall & Dale (1968a)	
P. nudum	Wall & Dale (1968a)	*Multispinula quanta*

Marine species with organic-walled cysts	References	Paleontological name for cyst
P. oblongum	Wall & Dale (1968a, Plate 1, Fig. 23,24)	
P. oblongum	Wall & Dale (1968a, Plate 1, Fig. 25–28)	*Votadinium calvum*
P. pentagonum	Wall & Dale (1968a)	*Trinovantedinium capitatum*
P. punctulatum	Wall & Dale (1968a)	
P. subinerme	Wall & Dale (1968a)	
Protoperidinium sp. 1	Wall & Dale (1968a)	
Protoperidinium sp.	Dale (unpublished) (Fig. 19,20)	*Brigatedinium majusculum*
Protoperidinium cf. *divergens*	Dale (unpublished) (Fig. 17,18)	*Peridinium ponticum*
Pyrodinium bahamense	Wall & Dale (1969)	*Hemicystodinium zoharyi*
Pyrophacus horologium	Wall & Dale (1971)	
P. vancampoae	Wall & Dale (1971)	*Tuberculodinium vancampoae*
Polykrikos schwartzii	Dale (1976)	
P. kofoidi	Morey-Gaines & Ruse (1980)	
Marine species with calcareous cysts		
Ensiculifera sp. cf. *mexicana*	Wall & Dale (1971)	
Scrippsiella sweenyae	Wall & Dale (1968b)	
S. trochoidea	Wall & Dale (1968b)	
Freshwater species with organic-walled cysts		
Ceratium hirundinella	Wall & Evitt (1975)	
C. horridum	von Stosch (1972)	
Gymnodinium pseudopalustre	von Stosch (1973)	
Peridinium cinctum fa. *ovoplanum*	Pfiester (1975)	
P. cinctum fa. Westii	Erėn (1969)	
P. gatunense	Pfiester (1977)	
P. inconspicuum	Wall, et al. (1973)	
P. limbatum	Wall & Dale (1968a)	
P. volzii	Pfiester & Skvarla (1979)	
P. willei	Pfiester (1976)	
P. wisconsinense	Wall & Dale (1968a)	
Woloszynskia apiculata	von Stosch (1973)	
W. tylota	Bibby & Dodge (1972)	

Other suface features on cysts that may be diagnostic include the reflected features already discussed. Paracingulum and parasulcus are most commonly represented, whereas flagellar pores, apical pores, trichocyst pores, and even growth bands may occasionally be reflected to some extent (Gocht, 1979).

Paratabulation. Biologist have informally recognized two groups of dinoflagellates based on their appearance in light microscopy, thecate (armored) and naked (unarmored) dinoflagellates. In armored dinoflagellates, the outer wall (theca) is seen to include transverse rows of usually four- to six-sided plates (see Fig. 1), whereas in unarmored dinoflagellates, comparable plates are seen in the light microscope only after special staining or in electron microscopy. The arrangement of thecal plates (plate pattern or tabulation) is one of the main taxonomic criteria used in the biological classification of dinoflagellates. A system of consecutively numbering plates, proposed by Kofoid (1907, 1909, 1911), has traditionally been used to describe tabulation.

Paratabulation is tabulation (i.e., plate patterns) reflected on cysts. Exceptional fossil cysts such as *P. pyrophorum* show remarkable resemblance to thecae, but the structures are fundamentally different. Thecae are made up of discrete cellulosic plates, each formed within its own vesicle with a weaker zone (suture) between. The plates are thus easily separated, for example, by a dilute sodium hypochlorite solution, a useful technique employed by taxonomists studying plate details. By contrast, paraplates on cysts seem to be defined by surficial features only (e.g., parasutural ridges). With the exception of certain parts of the archeopyle (discussed next), paraplates cannot be separated into discrete plates. Fundamental differences between cysts and thecae were particularly well illustrated in detailed studies by Gocht and Netzel (1976). With the help of scanning electron microscopy, they compared tabulation details of thecae in *Protoperidinium conicum* (Gran) Balech with paratabulation in fossil cysts of *P. pyrophorum.* The superficial resemblance between the two was shown to be largely due to tabulation details similar to those seen on the *external* thecal surface reflected in replica and in opposite relief on the *internal* surface of the cyst wall. These observations further supported the case for genetic control of morphogenesis in both cysts and thecae.

Paratabulation is expressed morphologically by a wide variety of surficial features on the cyst wall, well illustrated from the fossil record by Evitt (1969). In general, this involves distribution of sculptural elements such as ridges, processes, or smaller ornamentation. Paraplates may be outlined either by a concentration (see Fig. 1) or conspicuous absence of such elements along parasutural zones; alternatively, they may be represented by the number and relative positions of such elements, with no direct indication of parasutures. Paratabulation is usually incomplete compared with tabulation of the theca; particularly, smaller sulcal and cingular plates are often absent or only partially reflected.

Archeopyle. Evitt (1961) introduced the term *archeopyle* for the opening in a cyst through which excystment occurs. Usually the shape of the archeopyle suggests that it comprises one or more paraplates (e.g., Figs. 1, 21, 22, 36, 37). If one paraplate is involved, this is termed the *operculum* (e.g., Figs. 21, 22); more than one with sutures between are termed *opercular pieces* (e.g., Fig. 37). The archeopyle sutures open completely in some species, releasing the operculum or opercular pieces; in other species, they remain partially attached (e.g., Figs. 12, 15, 16). An alphanumeric system (the archeopyle formula) is often used in palynology to summarize archeopyle details (see Evitt, 1967b, and suggested improvements by Lentin & Williams, 1975, and Norris, 1978).

Until fairly recently archeopyles were thought to be restricted to the epicyst (i.e., anterior to the paracingulum), but it is increasingly evident that some occur on the hypocyst (e.g., *Pyrophacus vancampoae* (Rossignol) Wall et Dale, Wall & Dale, 1971; and *Peridinities* species, Dale, 1978) or even the paracingulum (*Nannoceratopsis* species, Piel & Evitt, 1980). Gocht and Netzel (1976) showed that the position of the archeopyle in some cysts is probably related to the system of overlap in corresponding thecal plates. Some thecal plates (keystone plates) overlap all or almost all adjoining plates and therefore probably are more easily dissociated, whereas other plates overlap some adjoining plates producing inherent lines of relative structural weakness. Boltovskoy (1973) described ecdysis in a freshwater dinoflagellate in which the theca opened by an "archeopyle" closely resembling the archeopyle of its cyst. Gocht and Netzel's work suggest this probably represents lines of relative structural weakness in thecal plate overlap reflected by the cyst archeopyle. However, in almost all dinoflagellates for which thecae and cysts are known, thecae do not routinely break open along sutures corresponding to the archeopyle sutures of the cyst. Therefore, the term *archeopyle* should be restricted to the cyst.

The archeopyle is one of the main taxonomic criteria used in the cyst-based classification system used in paleontology (discussed in 4.4 under The paleontological classification of dinoflagellates). Biologically, it is difficult to assign taxonomic rank to such criteria as yet, but incubation experiments with living cysts suggest the following general conclusions:

1. Cysts with the same type of archeopyle are probably closely related. For example, even morphologically very different cysts with the same type of archeopyle on incubation produce obviously similar dinoflagellates (Fig. 45), whereas cysts with markedly different archeopyles produced different dinoflagellates.

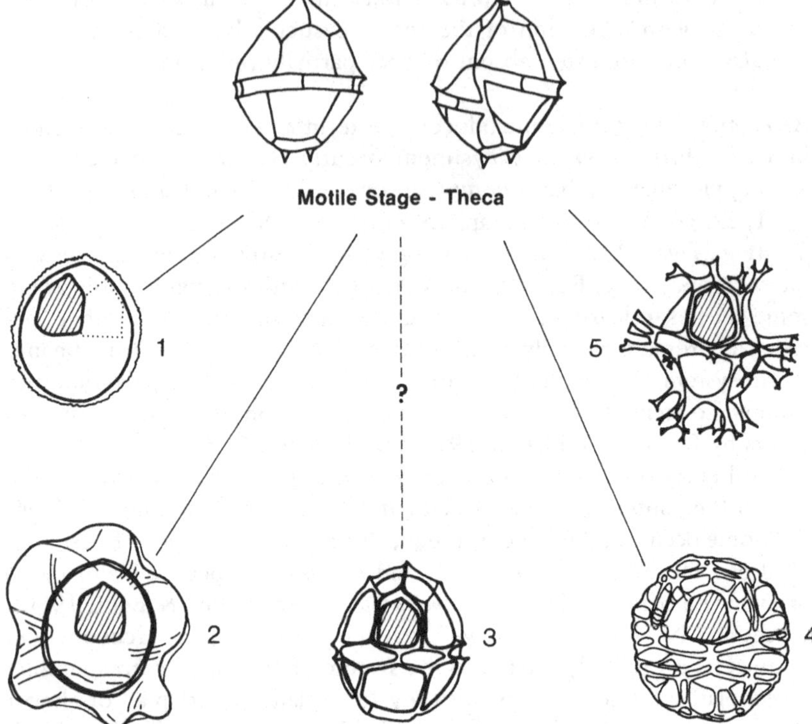

Fig. 45. Diagram illustrating the wide range of cyst morphology found within the *Gonyaulax spinifera* group of dinoflagellates. Motile stages from cysts 1, 2, 4, and 5 are known to share the same basic thecal morphology illustrated; cyst 3 is presumed to. Cyst types illustrated represent the following paleontological based "cyst genera": 1, *Tectatodinium* (1p archeopyle – several species) or *Bitectatodinium* (2p archeopyle – one species); 2, *Planinosphaeridium* (one species); 3, *Impagidinium* (several species); 4, *Nematosphaeropsis* (several species); 5, *Spiniferites* (both 1p and 2p archeopyles – many species).

2. Cysts with archeopyle differences involving only adjacent paraplates in the same series are probably also closely related (e.g., Figs. 36, 37, 45).

3. Cysts with markedly different archeopyles (i.e., involving very different paraplates) are probably not closely related (e.g., cysts of *Gonyaulax* species with archeopyles involving precingular paraplates compared with intercalary archeopyles in *Protoperidinium* species).

Incubation experiments have confirmed previously suggested limitations in the use of archeopyles in defining cysts. Paleontologists, having defined the archeopyle as the excystment aperture of dinoflagellate cysts, recognize archeopyles only in fossils showing cyst features. In practice, either direct evidence of paratabulation (since few cysts show other definitive features) is required or the paraplatelike shape of the archeopyle alone is often taken as evidence of paratabulation and used to differentiate cysts from acritarchs. Lister (1970) suggested this differentiation probably biased the fossil record toward armored dinoflagellates, since unarmored forms lacking tabulation in their motile stages might be expected to produce cysts similarly lacking paratabulation (including paraplatelike archeopyles). The few known cysts of unarmored dinoflagellates confirm this (see Figs. 23–25 and Wall & Dale, 1968a, p. 284). Lacking paraplate-shaped archeopyles and other "definitive" cyst features, such cysts would be classified as acritarchs in the fossil record. Furthermore, at least one armored dinoflagellate, *P. faeroense,* produces an acritarchous cyst in which the simple split of the archeopyle suggests no hint of paratabulation.

Cell contents. Little is known about the cell contents of cysts which are not easily studied. The resistant cyst wall is generally impervious to fixing and staining agents usually used in light microscopy and transmission electron microscopy. However, Bibby and Dodge (1972) successfully fixed, sectioned, and stained a small proportion of a sample of cysts from the freshwater species *W. tylota*, providing the only detailed study of cyst contents. Using both light microscopy and electron microscopy, they showed that, compared with motile stages, the main features of cyst contents included "the reduction in size or disappearance of cytoplasmic structures such as chloroplasts, Golgi bodies and pusule; and the enlargement of a central accumulation body and cytoplasmic vacuoles containing crystals." More recently, Dürr (1979) demonstrated the use of freeze-etching techniques for studying cyst contents.

As seen live through the cyst wall (where this is transparent enough) under normal light microscopy, cell contents may change dramatically according to whether the cyst is freshly formed, "mature," or ready to excyst (described in section 4.3 under The resting period and dormancy). Particularly in freshly formed cysts or those ready to excyst, the microgranular cytoplasm may develop rapid Brownianlike motion, and cell contents become noticeably darker. Mature cysts often include food storage products such as starch grains or oil droplets (Figs. 8, 9, 11). Some cysts (all known so far are autotrophic) include one or more conspicuous yellow- to red-pigmented bodies (around 2–5 μm in diameter) of unknown function (e.g., Figs. 5, 7, 44). These have been called *eye spots* or accumulation bodies by various authors without further clarification. Yentsch, et al. (1980, p. 253) reported that such bodies in probably fairly newly formed *G. excavata* cysts emitted a bright red, chlorophyll-like fluorescence when excited by blue light. A relatively large nucleus is sometimes visible in cysts, and this may be seen to revolve within the cyst for a short period presumably just prior to meiosis (nuclear cyclosis, described by von Stosch, 1972).

4.3 Cysts and dinoflagellate life cycles

Work by paleontologists in the past 15 years has shown that many living dinoflagellate life cycles include cysts. These investigations were more concerned with the paleontological significance of living cysts, and many questions concerning the biological processes involved remain unanswered. Current knowledge of encystment, dormancy, excystment, and the probable role of cysts in sexual life cycles is summarized here. Clearly, cyst production is a much more widespread and important process than the phycological literature of the past has suggested. However, much of the basic information, though often repeated, particularly in the paleontological literature, is at best based on very few observations. Systematic phycological investigation of these processes is overdue.

Encystment

Which dinoflagellates form cysts? How representative of "real plankton" is the "benthic plankton," the cyst assemblage in bottom sediments? At least 170 distinctive cyst morphotypes have been seen in Recent sediments, many with live cell contents; sediments from previously uninvestigated regions still often yield new types, suggesting that probably well over 200 cyst types are produced today. About 50 of these have been correlated with their dinoflagellate motile stages (see Table 2) through observations of encysting plankton, encysting laboratory cultures, or cyst incubation experiments. A striking feature of these data is the high proportion of cysts not only belonging to just two genera, *Gonyaulax* and *Protoperidinium*, but mainly restricted to only a few groups within these genera. Although cysts are known from at least some species of other genera (see Table 2), it seems likely that few or no cysts are produced within many common genera. For example, cysts are known only for freshwater species of *Ceratium*, whereas no cysts are known for species of *Dinophysis* Ehrenberg or *Prorocentrum* Ehrenberg though sediments often have been examined from areas where these genera are common in plankton. Dale (1976) found 26 cyst types in bottom sediments from a Norwegian fjord where 57 species of motile dinoflagellates were recorded in long-term plankton records. Although this is probably fairly representative of temperate neritic plankton, it is almost certain that a much smaller proportion of cyst-forming species is typical for tropical waters and the open ocean (Dale, in preparation).

Where do cysts form? Living cysts have been recovered from bottom sediments in all major aquatic environments (freshwater lakes, brackish neritic water, and fully marine open ocean) and climatic/biogeographical zones from tropical to arctic regions particularly in the North Atlantic (see Fig. 46, and Wall et al., 1977). Cysts have also been recorded occasionally from the water column during encystment of planktonic

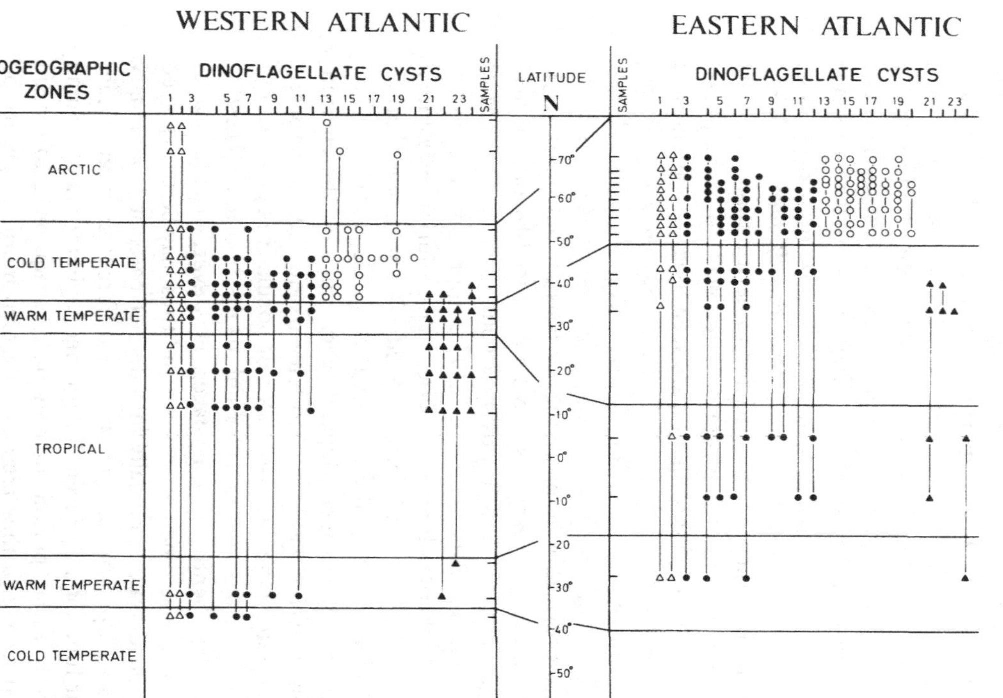

Fig. 46. Distribution of selected living dinoflagellate cyst types compared with standard biogeographical zones for the Atlantic. Living cysts (numbered 1–24) were recorded from bottom sediments or plankton in coastal regions from both sides of the Atlantic. Data are based on more than 200 samples (from Wall et al., 1977; Dale, unpublished), which are clustered into regions for ease of presentation. Cysts: 1, *O. Centrocarpum;* 2, Undifferentiated *Protoperidinium;* 3, *S. bulloideus;* 4, *P. conicum;* 5, *L. Machaerophorum;* 6, *Polykrikos;* 7, *S. mirabilis;* 8, *P. pentagonum;* 9, *P. oblongum;* 10, *V. calvum;* 11, *P. claudicans;* 12, *P. avellana;* 13, *P. conicoides;* 14, *S. elongatus;* 15, *S. membranaceus;* 16, *P. denticulatum;* 17, *B. tepikiense;* 18, *Protoperidinium sp.;* 19, *P. faeroense;* 20, *P. membranaceum;* 21, *P. vancampoae;* 22, *H. zobaryi;* 23, *O. israelianum;* 24, *P. subinerme.*

populations (e.g., Fig. 4) or as a result of resuspension from bottom sediments. Encystment has been reported several times from natural environments (e.g., Nordli, 1951; Wall & Dale, 1968a; Bibby & Dodge, 1972), and I have observed this regularly in the Oslofjord, Norway, for the past 5 years (mostly unpublished observations). A few species have encysted spontaneously in cultures (e.g., *Scrippsiella trochoidea* (Stein) Loeblich III: Braarud, 1945; Wall et al., 1970; *G. grindleyi, P. reticulatum*: Braarud, 1945; Dale, unpublished observations; *P. faeroense*: Dale, 1977a). The marine evidence, so far restricted to neritic zones, suggests that encystment begins within the plankton (often well up in the photic zone) where early stages of cyst formation are seen together with normal vegetative motile stages. These early stages may be recognized by immature ornamentation (e.g., rudimentary processes as in Figs. 41, 42) and cell contents, and they may retain all or part of the enclosing parent theca (Figs. 41, 42). Cyst formation in many species lasts for only a few days. This, together with the fact that cysts may be produced by only a small proportion of the motile stages, probably accounts for the paucity of reported encystment in plankton records. The approximate ratios of cysts to motile stage thecae observed at Woods Hole during encystment were 1 : 2 for *G. grindleyi,* 1 : 120 for *Protoperidinium oblongum* (Aurivillius) Parke et Dodge, and 1 : 500 for *G. digitalis* (Dale, 1976, p. 53).

The fate of cysts, once formed, depends very much on the environment. In shallower neritic water, cysts often sink quickly in the water column to concentrate in the unconsolidated or floculative upper layer of bottom sediment. In cultures, too, cysts sink to the bottom, suggesting that basically they are benthic resting stages. However, cysts behave as fine silt particles in the sedimentary regime and thus may be transported by currents or concentrated gradually into deeper parts of sedimentary basins. Presumably open-ocean cysts have a different strategy, either completing their function before sinking too deeply in the water column (e.g., by a shorter resting period) or they may be equipped to remain within a depth from which they are able to return motile stages to the plankton after excystment (e.g., by density regulation).

When do cysts form? Unfortunately, observations of cyst formation are restricted almost entirely to temperate and north-temperate regions (e.g., Woods Hole, northeast U.S., and Oslofjord, Norway). Not surprisingly, cyst formation in these regions is markedly seasonal, closely following the well-known seasonal succession of motile stages in the plankton. Most fall into one of two well-defined categories: species that reach their maximum abundance in spring plankton and form cysts in late spring/early summer (e.g., *P. oblongum,* Wall & Dale, 1968a; *P. faeroense,* Dale, 1977a) and species with maximum abundances in summer that form cysts in late summer/autumn (e.g., *G. digitalis*, Wall & Dale, 1968a; *G. polyedra*, Nordli, 1951; Dale, unpublished; and *P. horologium*, Dale, unpublished). There may be at least two other categories: species with two annual peaks of maximum abundances, typically in spring and autumn, and cyst formation following each peak (e.g., *G. tamarensis*, Anderson & Morel, 1979; *G. grindleyi*, Dale, unpublished) and species with sporadic abundances and cyst formation spread over spring, summer, and autumn (e.g., *S. trochoidea*, Dale, unpublished). The "double" cyst formation noted for *G. grindleyi* in Norwegian fjords probably contributes to the relative overrepresentation of its cysts in sediments discussed by Dale (1976, p. 53). To understand the process of cyst formation, it will be necessary to obtain data comparable to that summarized previously from tropical waters less influenced by seasonal plankton blooms.

Why are cysts formed? What are the factors inducing cyst formation? As yet, there is insufficient evidence to properly answer these questions, but relevant observations and suggested interpretations are discussed here. Wall and Dale (1968a, Fig. 2) illustrated the alternation of motile stages and cysts of *G. digitalis* in coastal waters at Woods Hole, N. E. U.S. They suggested that cyst formation in this case served an obvious overwintering function during the period when water temperatures are too cold for survival of the motile stages, comparable with the life cycles of many temperate freshwater dinoflagellates (e.g., *C. hirundinella*, Huber & Nipkow, 1922, 1923). This analogy has been taken much further by others who presumed that *G. digitalis* cysts that formed in late August/September (Wall & Dale 1968a, Figs. 2, 3) did so in response to a fall in seawater temperature; this is often cited to support the theory that cyst formation is induced by the onset of adverse conditions (e.g., Sarjeant, 1974, pp. 43–44). In many ways, this is a plausible theory; certainly "reversing adverse temperatures" by returning cysts to a favorable temperature has long been used to induce excystment in incubation experiments (e.g., Huber & Nipkow, 1922, 1923) and was used by Wall and Dale. However, although temperature data were not reported by Wall and Dale (1968a), such data were recorded and clearly show that *G. digitalis* encysted in Woods Hole waters before seawater temperature began to fall after the summer. Furthermore, several attempts at cooling laboratory cultures of known cyst-forming dinoflagellates at rates comparable to natural conditions, and many more erratic accidental "cooling experiments," caused by mechanical failures in culture facilities, all failed to trigger encystment (Dale, unpublished). Obviously, temperature as one possible factor inducing encystment cannot be eliminated without a thorough investigation of encysting natural plankton and cultures, but as yet I know of no evidence for temperature-controlled encystment.

There is rapidly accumulating evidence that lowering nitrogen levels in cultures may induce sexuality, which in some species involves cyst formation (e.g., von Stosch, 1973; Pfiester, 1975, 1976, 1977; Walker & Steidinger, 1979). Sexuality and the role of cysts is discussed at the end of this section, but of interest here is lowered nitrogen as a factor inducing encystment. As with most culture work, the most important question "how does this relate to natural conditions?" is very difficult to answer. I know of no field data directly relating cyst formation to lowered nitrogen levels and the little circumstantial evidence available suggests that, although this could possibly cause some encystment, other factors are sometimes almost certainly involved. Observations by Wall and Dale (1968a, Fig. 2) suggest that for a given species, encystment often begins around the peak of exponential growth and maximum abundance of its motile stages in the plankton. Sometimes these are large blooms, which could well deplete nitrogen levels significantly, but more often they account for only a very small part of the total phytoplankton. Similarly, although some encysting cultures were old or contained dense concentrations of motile cells (e.g., Braarud, 1945; Wall et al., 1970; Dale, 1977a) suggesting the strong possibility of lowered nitrogen levels, fresh cultures have encysted on occasion, suggesting that other factors are involved (Dale, 1977a, and unpublished). The extent to which encystment is simply a response to "adverse conditions," as often presented, or is a routine stage in the life cycle, favored rather than inhibited by optimal conditions for vegetative growth, remains unproven.

How are cysts formed within motile stages? A discussion of encystment would be incomplete without reference to the "contractional-growth" hypothesis for cyst formation. This was developed in the geological literature in the early 1960s and has been repeated many times since (e.g., Evitt, 1969; Sarjeant, 1974; Williams, 1978; Brasier, 1980). Evitt (1961) originally hypothesized that some fossil cysts with processes had originally developed inside a nonfossilizable theca, later illustrating this with a drawing of a cyst with long, fully developed processes inside an intact theca (Evitt, 1963, Fig. 3B). Evitt and Davidson (1964) reported confirmation of this type of cyst formation which they observed in preserved modern plankton, illustrating this with a photograph of a cyst with well-developed processes and two of the original thecal plates still attached (their Plate 1.10), and a diagram of a similar cyst inside an intact theca (their Fig. 2). Sarjeant (1965) and Sarjeant and Downie (1966, p. 13) took this theory a stage further, recognizing two different groups of cysts based on the degree of supposed contraction of the cyst body from the theca during cyst formation.

Wall (1971, p. 4) pointed out that since reports up to that time had figured only remains of part of a theca surrounding such cysts, there was no evidence that cysts had grown by contraction from the theca, and he stated, "there is clearly a possibility that the thecal lumen is fully occupied by the cyst before any spine growth takes place and that later, at maturity, the overall size of the cyst may exceed that of the parental theca."

My own observations support Wall's comments on the contractional growth hypothesis. During encystment of "spiny" cysts of *G. grindleyi, G. polyedra, P. faeroense, G. digitalis,* and *G. spinifera*, both in the Oslofjord and in cultures, I routinely see either newly forming cysts with immature processes fully occupying the thecal lumen, or cysts with fully developed processes but without thecae or at most with only parts of thecae still attached (e.g., Figs. 41–44; Dale, 1977a, Fig. 24). Thecal plates attached to fully developed cysts usually show no evidence of decay, and intact empty thecae of normal vegetative stages are commonly seen in the same samples. I therefore suggest that cysts with relatively long processes (chorate cysts) form not by contraction of the cyst body as the processes grow but with process growth actively helping to break up the theca by an overall expansion of the cyst (see Fig. 3a). One way to test this is to compare overall sizes of cysts and thecae of planozygotes producing them (not thecae of normal vegetative cells, which usually are smaller than those of planozygotes, see discussion in section 4.3 under Cysts and sexual cycles). In the one available report of such measurements (Dale, 1977a, p. 245), overall cyst size clearly exceeded that of the theca.

The resting period and dormancy

A great deal of evidence has now accumulated suggesting that encystment is followed by a resting period, the first part of which is mandatory (dormancy). Almost all the evidence is from temperate regions, and much of it concerns freshwater species. The well-documented overwintering of freshwater cysts in these regions involves three main phases: encystment toward the end of summer plankton, resting of benthic cysts during winter, and excystment in spring regenerating the plankton (e.g., *C. hirundinella* in Lake Zurich, Huber & Nipkow, 1922, 1923). Wall and Dale (1968a) documented a similar cycle for some marine dinoflagellates at Woods Hole. In these and many other studies, at least part of the resting period was considered mandatory, since freshly formed cysts could not be made to excyst by temperature manipulations known to cause excystment later on.

Though the apparent need for a resting period and dormancy in cysts has been known for over 50 years, we know little regarding the details. Available data are mainly concerned with the amount of time spent in resting and dormancy and how this varies with respect to other factors, especially temperature. Good estimates have been made for the natural resting period in temperate lakes and coastal waters by monitoring motile stages in plankton, cyst formation in plankton, cysts in bottom sediments, and their viability. In these strongly seasonal environments, cyst formation in many species appears to be confined to just a few days, and this fresh supply of cysts has permitted experiments to estimate the effects that factors, such as temperature, have on resting period and dormancy. Results show that in temperate and north-temperate environments, many dinoflagellates spend much more time as resting benthic cysts than as motile plankton. For example, estimated time per year spent encysted was 7–8 months for *C. hirundinella* in Lake Zurich (Huber & Nipkow 1922, 1923), 9 months for *G. digitalis* at Woods Hole (Wall & Dale, 1968a, p. 268), and 10 months for *P. horologium* in the Oslofjord (Dale, unpublished). Estimated mandatory dormancy for *C. hirundinella* and *G. digitalis* in the above studies were 6 weeks and 3 months, respectively. However, Anderson (1980) has shown that such estimates of dormancy may be expected to vary according to temperature. He found that in freshly formed cysts of *G. tamarensis* stored at 22°C, the required dormancy was probably less than 6 weeks, while this was increased to about 4 months in similar cysts stored at 5°C.

Termination of the resting period by excystment is retarded by low temperatures (7°C) commonly found in temperate waters during winter or elsewhere in deeper waters. For example, Huber and Nipkow (1922) reported this response (at 5°C) for *C. hirundinella*, excystment of which was also retarded at higher temperatures (25°–30°c). While Pfiester (1975), in one of the few studies reported so far using cysts produced in cultures and therefore of a more precisely known age, showed that cysts of *P. cinctum* excysted after 7–8 weeks at 20°C, but excystment was retarded for at least 5 months at 4°C. However, shorter periods of "cold treatment" reportedly shorten dormancy in *P. cinctum* (Dürr, 1979). That excystment is retarded by local winter temperatures, rather than just low temperatures, is suggested by the response of *Pyrodinium bahamense* Plate cysts. Wall and Dale (1969) reported cysts of this species collected in Bermuda in February 1968 and stored at winter temperatures of approximately 16°C for at least 6 months without excysting, though they readily excysted at 26°C in March/April.

Under unfavorable conditions for excystment (e.g., low temperatures), cysts may remain viable for several years. Huber and Nipkow (1923), sampling successive varves in bottom sediments, estimated that at bottom water temperatures around 5°C in Lake Zurich, cysts of *C. hirundinella* remained viable for 6.5 years, while *P. cinctum* cysts remained viable for 16.5 years. Wall and Dale (1969, p. 147) discuss the possibility that this difference is because cyst walls in *C. hirundinella* are thinner than those of *P. cinctum*. Wall (1971, p. 11) mentioned cysts of *P. bahamenese* and *G. digitalis* that were viable after storage for at least 12 months at 10°C, and I have incubated and successfully germinated common cyst types [including *Gonyaulax spinifera* (Claparède et Lachmann) Diesing, and *G. grindleyi*] from Oslofjord sediment stored at 4°–7°C for 5.5 years.

Although controlled experimental results are lacking, there is evidence suggesting that cysts are able to withstand various adverse environmental conditions. Anoxic conditions seem to pose no threat to the cysts during normal resting periods. Many of the sediment samples from which I have recovered viable cysts were stinking black muds lacking evidence of benthic animals, and I routinely concentrate viable cysts from plankton tows by allowing these to decay for months in a refrigerator at temperatures around 5°–7°C. The latter method exposes cysts to both enormous bacterial and fungal growth and eventually anoxic, H_2S-rich waters. Huber and Nipkow (1923) reported destruction of *C. hirundinella* cysts by freezing or dehydration, and I have seen individual cysts that died as a result of accidental freezing during incubation experiments or drying out on microscope slides under observation. However, under natural conditions, it is possible that bottom sediment (especially muds or clays) would not dry out completely for a long time even if exposed to air and cysts could remain viable with the help of trapped pore water. Similarly, the properties of sediment exposed to freezing conditions may reduce the adverse effects for cysts buried within.

Very little is known regarding the cellular activity of cysts during the resting period. It seems reasonable to suppose that metabolic activity is minimal in cysts able to remain viable for many years under conditions unfavorable for excystment. (Bibby & Dodge, 1972, p. 97, show evidence for this.) Cell contents, however, change dramatically at times suggesting accelerated activity. Many reports mention changes in cyst contents during the resting periods (e.g., Wall & Dale, 1969; von Stosch, 1973; Pfiester, 1976; Dale, 1979; Anderson, 1980; Yentsch et al., 1980). The general sequence of changes observed may be summarized:

1. *An initial transition stage where typical motile stage contents change to contents more characteristic of cysts.* This often involves visible Brownian-like motion in the cytoplasmic particles (often within distinct vacuoles) and darkening of the overall contents. Depending on species, these changes may be well under way in the motile cell (planozygote) before obvious cyst formation, or they may begin after loss of motility.
2. *Visibly stable cell contents typical of the resting period.* In almost all cases, Brownian-like motion is no longer visible (*G. polyedra* is a notable exception), and cell contents include large amounts of storage products such as starch grains or oil globules; yellow- to red-pigmented bodies may be prominent in some species. I have seen cell contents remain at this stage without visible change for more than 12 months in controls of incubation experiments.
3. *A transitional stage that visibly reverses changes seen in stage 1.* Contents often darken intensely as rapid Brownian-like motion proceeds again, and eventually (either before or after excystment) typical motile cell contents are restored.

The darkening of cell contents in stages 1 and 3 is particularly visible in autotrophic species and may represent photosynthetic pigments

(Yentsch et al., 1980); in many nonphotosynthetic *Protoperidinium* cysts, cell contents remain pale throughout, and some show no obvious changes prior to excystment.

Excystment

Excystment (sometimes called *germination*) has usually been induced in culture experiments by raising the ambient temperature (e.g., Huber & Nipkow, 1923; Wall & Dale, 1968a; von Stosch, 1973). Exceptions to this include studies by Anderson (1980) in which *G. tamarensis* cysts excysted after lowering the temperature from 22° to 16°C, and by Pfiester (1975) in which *P. cinctum* cysts excysted after 6–7 weeks at the same temperature (20°C). Almost all excystment data are from temperate regions and are therefore possibly influenced by seasonality. In many experiments, excystment was induced by manipulating ambient temperatures around those encountered in nature. Typically, cysts were isolated from bottom sediments in late winter/early spring (presumably having completed dormancy) and exposed to spring/summer temperatures in culture experiments. Alternatively, newly formed cysts from nature or cultures were stored in the laboratory at winter temperatures before exposure to higher temperatures favorable for excystment. For most temperate species, "winter" temperatures used were <10°C, and "summer" temperatures, 15°–26°C (e.g., experiments of Wall & Dale, 1968a). For subtropical/tropical species, these temperatures were 15°–25°C and >25°C, respectively (e.g., experiments of Wall & Dale, 1969, from Bermuda and Puerto Rico, and those of Erėn, 1969, from a lake in Israel).

Available data strongly suggest the possibility that temperature is a major factor inducing excystment in the natural environment, at least in temperate regions. There is generally close correspondence between experimental data concerning excystment as a response to temperature changes and those that would be required in nature. The possible roles in excystment of other factors, such as day length and water chemistry, have not been systematically investigated, though these are thought to have little or no effect on other stages such as encystment (von Stosch, 1973) and dormancy (Anderson, 1980). Excystment seems to proceed equally well both in light and total darkness (Dale, unpublished).

However, temperature as a factor "inducing" excystment should obviously be considered as part of the basic time/temperature relationship involved also in the preceding stages of dormancy and resting, a relationship that is still not understood. Huber and Nipkow (1922, 1923) provide some of the few experimental data concerning part of this, but other parts need clarifying if we are to understand the phenomenon as a whole. This is one of the more important areas for biological research on cysts, because the physiological properties of cysts may be more responsible for determining ecological limits to where and how a given species lives (discussed in section 4.5 under Probable functions of cysts) than similar properties of their motile stages. Important fundamental principles for using fossil cysts as paleoclimatic indicators are also bound up in the same questions. The main points to be resolved include:

1. *Concerning dormancy.* Is this a maturation phase, the duration of which is a simple function of time and temperature? For example, do the data of Anderson (1980, Figs. 17A, B) represent maturation (dormancy) in cysts stored at 5°C taking approximately threefold longer than at 22°C, because this was an approximately threefold higher temperature (100–110 versus 30–40 days, respectively)? If the only other available datum for *G. tamarensis* (8 weeks dormancy at 17°C from Turpin et al., 1978) is accepted as a possible intermediate value, a linear relationship between dormancy, time, and temperature is suggested. Is the age of the cyst in itself an important factor, as reported for the freshwater species by Huber and Nipkow (1922, 1923), such that excystment is inhibited for several weeks after formation but then becomes progressively more rapid toward an optimum time for excystment? Is there one basic time/temperature relationship governing dormancy in cysts, or are we already seeing evidence for at least two different mechanisms, with more "flexible" dormancy requirements in *G. tamarensis* complementing its basic strategy as a bloom species exploiting the changeable environments in estuaries and coastal embayments? To what extent may dormancy be broken by shock treatments? In many experiments so far, large, rapid changes of temperature and sometimes sonication have been used; we need to know experimental limits for their use, beyond which they may unduly influence results.

2. *Concerning a resting period after dormancy.* Does this occur only if cysts are maintained at temperatures inhibiting excystment, or is there possibly an afterripening effect comparable with the annual biological rhythms that Yentsch and Mague (1980) believe effect motile stage growth in *G. tamarensis*?

3. *Concerning excystment.* Huber and Nipkow (1923, Fig. 11) showed the process of excystment as a simple function of time and temperature, with excystment in *C. hirundinella* completed in less than 36 hr at an optimum temperature of 21°–22°C and taking progressively longer time both at higher and lower temperatures up to inhibition levels. Wall and Dale (1969, p. 146) noted a similar tendency in *P. bahamense* cysts from Bermuda, and the whole process may take from less than 2 days to several weeks depending on temperature. Thus, for a given species there is probably a "temperature window" within which it may excyst, time allowing, and outside of which excystment is inhibited; but how and why excystment happens therefore depends very much on the processes of

dormancy and resting already discussed. For example, if dormancy is completed at a temperature within the excystment window, then in the absence of a need for further "ripening," excystment will occur as a natural consequence of the end of dormancy without the need for a stimulus (e.g., *P. cinctum* after 7 weeks at 20°C in experiments of Pfiester, 1975, or *G. tamarensis* after 8 weeks at 17°C, Turpin et al., 1978). This may well account for excystment in *P. bahamense* and other species in tropical waters where motile stages are present in plankton year-round and temperature varies by only 1° or 2°C. The concept of needing a stimulus for excystment appears often in the literature, but this may be misleading. Even the classic overwintering cysts of temperate waters may *require* no particular stimulus, excysting in spring simply when returned to the temperature window. In this case, any additional "resting period" may be real inhibition of excystment, or excystment may have already begun but takes a long time at temperatures toward the outer limits of the window. Further studies should examine the extent to which dormancy and excystment are really one time/temperature-related process (e.g., part of the nuclear exchange in sexual reproduction, discussed next), which may be modified by or show adaptations to the wide fluctuations in temperatures encountered in many environments (e.g., by allowing reactions to proceed even after interruptions of up to several years).

The process of excystment as seen under light microscopy has been described and illustrated by several authors (e.g., *C. hirundinella*, Huber & Nipkow, 1922; *G. digitalis*, Wall, 1965; *G. spinifera*, Wall & Dale, 1968a, 1970; *G. tamarensis*, Anderson & Wall, 1978). Immediately prior to excystment (after changes discussed earlier in this section), cell contents often withdraw noticeably from the cyst wall, and the cingulum and other major features of body shape may begin to form. After the archeopyle opens, the protoplast (surrounded by a thick, hyaline gelatinous sheath) flows through the archeopyle by amoeboidlike motion. Under natural conditions, this probably takes only a few seconds to emerge, but experimental conditions may prolong this up to several hours. The longitudinal flagellum (which may sometimes be "doubled," see discussion in section 4.3 under Cysts and sexual cycles) may be formed just before excystment or usually within a few minutes after, followed shortly by development of the transverse flagellum. This stage has been referred to as a gymnodinioid stage, since even armored species require a day or so to form a theca (Fig. 43).

Cysts and sexual cycles

Until fairly recently dinoflagellates were not generally believed to reproduce sexually, though Zederbauer (1904) had reported meiosis in *C. hirundinella.* Work by von Stosch (e.g., 1964, 1965, 1972, 1973) was largely responsible for changing this impression, and in the past 15 years it has become increasingly obvious that sexuality in the dinoflagellates is widespread. Though sexual cycles have been documented for only 16–20 species so far, at least 10 are known to involve resting cysts. The most detailed genetic work on dinoflagellates has been concentrated on *Crypthecodinium cohnii* (Seligo) Chatton (e.g., Tuttle & Loeblich, 1974; Himes & Beam, 1975), but this and other species not forming resting cysts are not discussed here.

All known dinoflagellates except *Noctiluca miliaris* Suriray (Zingmark, 1970) are haplonts. Their known sexual cycles share basic features that include production of gametes, pairs of which fuse to produce a zygote and eventually meiosis. However, it is the zygote that is of particular interest here. There are basically three types of zygotes described so far: In some species the zygote (planozygote) is motile throughout, and meiosis is completed without cyst formation (e.g., marine *Ceratium horridum* Gran, von Stosch, 1972). In some other species, the planozygote loses motility and forms a temporary cyst (e.g., *Helgolandinium subglobosum* von Stosch, von Stosch, 1972). In a third group of species, the planozygote forms a resting cyst (hypnozygote). The cysts dealt with in this review are most likely hypnozygotes, though as yet this is only proved in a few species.

I previously suggested the strong probability that fossil cysts were hypnozygotes (Dale, 1976, pp. 56–57) and have since reported observations consistent with this for the fossilizable cyst of *P. faeroense* (Dale, 1977a). Depending on the species, living cyst walls show a variable degree of preservability, and possibly none of the hypnozygotes documented so far can fossilize. However, the cell contents and physiological properties of these cysts seem identical with those of "living fossil" cysts, suggesting that differences only involve the degree of preservability of the wall and that all share the same function as hypnozygotes in sexual cycles. Indeed, if fossil cyst walls include sporopollenin, this fact may point to a sexual function of cysts, since this material is only found elsewhere as a basic protection for sexual stages in plants (e.g., the walls of pollen grains and spores in higher plants, and spores of some algae and fungi; Brooks & Shaw, 1973).

Reexamination of my observations from earlier incubation experiments with living fossil cysts at Woods Hole (in light of subsequent work by von Stosch) revealed marked similarities between these and sexual cycles reported by von Stosch (e.g., occasional double longitudinal flagellae in the emerging protoplast), and I am confident that eventually these too will be shown to be hypnozygotes (Dale, 1976, p. 57). However, even if all dinoflagellates have sexual cycles of the types reported so far, large gaps in the "fossil record" being produced now would still result. Some species complete meiosis without cysts, some produce nonfossilizable temporary cysts or hypnozygotes, and only a minority of species produce fossilizable cysts (hypnozygotes). There is increasing awareness of probable large gaps in the fossil record of dinoflagellates (e.g., Piel & Evitt, 1980, p. 101). The comparison with living cysts suggests differences in the type of zygote produced as a plausible explanation (Dale, 1976, p. 57). Implications of this explanation for considerations of dinoflagellate evolution are discussed in section 4.5 under Cysts: a survival strategy?

The basic sexual cycle producing cysts as hypnozygotes is schematically illustrated in Fig. 32. As with so many other features of dinoflagellates, the details may differ markedly, and some of the important variations are summarized in Table 3. Most of the data so far are from von Stosch (1964, 1972, 1973) and more recently Pfiester (1975, 1976, 1977) and Pfiester and Skvarla (1979). The main features of gametes and zygotes in cyst-producing sexual cycles are summarized:

Gametes. Gametes described so far often are almost identical with the normal vegetative cell and have probably been confused with these in most plankton work. Even fusing gametes are thus easily confused with normal dividing vegetative cells, which (rather than the supposedly rare incidence of sexuality) probably accounts for the paucity of reported sexual cycles in dinoflagellates. Reported visible differences between gametes and vegetative cells include: size (gametes are usually smaller, especially in anisogamous species); swimming motion (gametes may swim noticeably faster and sometimes collect in "dancing" groups, within which pairing takes place); and cell contents (gametes are sometimes paler colored but may contain prominent yellow- to red-pigmented bodies as described for cysts). Gamete formation in at least some species occurs in the dark part of the light/dark cycle, and in some species (e.g., *Woloszynskia apiculata* von Stosch) this is reversible such that gametes return to normal vegetative cells if conditions are favorable, whereas in other species (e.g., *G. tamarensis*) this is irreversible. *W. apiculata* gametes are negatively phototactic. Gamete formation usually occurs in only a small part of the total population (e.g., if exposed to low nitrogen levels), and in at least one population the vegetative cells continued to increase in culture conditions that caused gamete formation in some cells (Pfiester, 1977).

Zygotes. Fusion of gametes usually gives rise to a motile zygote (planozygote), which initially often resembles the normal vegetative cell, with which it probably has been confused. However, the planozygote is usually noticeably larger than a normal vegetative cell and often swims more slowly. The planozygote is relatively long-lived as in the Volvocales (3–20 days) and may gradually enlarge or become grossly misshapen

Table 3. *Main variations in details of sexual cycles in dinoflagellates*

Species	Vegetative cell		Gametes		Hypnozygote		Excystment cells released	Meiosis	Reference
	Homothallic	Heterothallic	Isogamete	Anisogamete	Cyst	No cyst			
Ceratium cornutum		x		x	x		1 Meiocyte	Completed outside cyst	von Stosch (1972)
C. horridum	x			x		x		In planozygote	von Stosch (1972)
Gonyaulax monilata		?	x		x		1	?	Walker & Steidinger (1979)
Gymnodinium pseudoplaustre	x		x		x		1 Meiocyte	Completed outside cyst	von Stosch (1972)
Woloszynskia apiculata		x	x		x		2 Swarmers	I division in cyst, II division outside	von Stosch (1973)
Peridinium cinctum	x		x		x		1	? In cyst	Pfiester (1975)
P. gatunense	x		x		x		2	? Partially in cyst	Pfiester (1977)
P. volzii		x	x		x		1	?	Pfiester & Skvarla (1979)
P. willei	x		x		x		1	?	Pfiester (1977)

(warty or lumpy), with massive intercalary bands developed in some thecate forms (e.g., Pfiester & Skvarla, 1980). In many cases, the planozygote is propelled by double ("ski-track") longitudinal flagella. Observations from culture experiments show that just prior to cyst formation, the planozygote swims gradually more sluggishly, finally coming to rest on the bottom. Cell contents are often packed with starch at this stage (e.g., Figs. 39, 40). Meiosis is thought to take place either in the planozygote, which then excysts to release an appropriate number of daughter cells (two to four, depending whether meiosis is complete), or by subsequent divisions of the single motile protoplast (meiocyte) released by excystment.

4.4 Cysts and dinoflagellate classification

Ideally, both motile stages and cysts (where present) should be included in defining the basic taxonomic unit (holomorph) of a dinoflagellate classification. Wall and Dale (1968a, p. 292) suggested this, initiating a lively debate in the paleontological literature during the past decade (Evitt, 1970; Norris & McAndrews, 1970; Harland, 1971, 1974, 1977; Sarjeant & Downie, 1974; Reid, 1974; Dale, 1976, 1978; Reid & Harland 1977; Bradford, 1978). There is general agreement *in principle* on combining motile stage and cyst information (e.g., Evitt, 1970, p. 38: "From any biological perspective this seems as virtuous as motherhood"; also Reid & Harland, 1977, p. 151; Bradford, 1978, p. 195), but opinions vary widely on whether or not (or how actively) this goal should be pursued at present.

Currently there are two different classification systems for dinoflagellates that have developed almost independently for over 100 years; a biological classification largely based on nonfossilizable motile stages in plankton and a paleontological classification based on the morphology of fossil cysts. Attempts to correlate cysts and motile stages through incubation experiments with living cysts have forced us to examine these classifications more closely, revealing major weaknesses in both (Dale, 1978). Reconciling differences between the two systems is more than just following up the obvious biological ideal of including two stages in the life history of one organism in one classification; cyst and motile stage information uniquely complement each other such that combining them offers the possibility of a much better classification system. For this to be realized, however, biologists and paleontologists working with dinoflagellates will first have to understand each other's points of view. Some of the main points will be summarized concerning dinoflagellate classification considered from three points of view, the biological, the paleontological, and the special combination of both facing anyone working with Recent or living cysts.

The biological classification of dinoflagellates

There are major difficulties concerning the classification of microscopic organisms such as dinoflagellates, for a long time believed to lack the usual sexual basis of a biological species and which are not even obviously plants or animals. The extent and significance of cysts was only recently realized, so that biologists have developed a classification based largely upon the morphology of motile stages from plankton. The success of this has varied greatly, depending on the type of dinoflagellate and the equipment, patience, and skill of the taxonomist.

Some dinoflagellates are fairly easily distinguished by their characteristic body shapes (e.g., species of *Ceratium* or *Dinophysis*); these species were defined early, using obvious, major morphological criteria that are still used today, and such forms are regularly reported from plankton. Many thecate forms, however, are only properly distinguished after careful analysis of plate patterns (including very small sulcal and cingular plates); some of these species have been carefully described, but for many others the literature offers only incomplete descriptions, and records of such forms are open to doubt. Many unarmored forms are even more difficult to work with; they are almost impossible to fix or preserve intact, and soon disintegrate if exposed live to the rigors of examination under the microscope.

Dinoflagellate taxonomy is extremely demanding and time-consuming work. It is therefore not surprising to find that, despite the long history of well over a century during which dinoflagellates have been studied, few have specialized in their taxonomy, and the enormous task of classification remains far from being adequately completed. Painstaking work by specialists such as Kofoid, Graham, or more recently Balech has shown that probably relatively small morphological features (e.g., sulcal and cingular plates) are very important for distinguishing taxa. However, most dinoflagellates have never been examined in such detail. A state of taxonomic imbalance therefore exists at present, whereby some dinoflagellates are well classified using minute details of minor features while many others are poorly classified using only major features that even then may be inadequately described. One of the main objectives of dinoflagellate taxonomy has to be a thorough reexamination of minor features in those forms not yet reported in such detail. Where possible this should combine culture work and observations from plankton, so that natural variation within a species is considered in assessing the taxonomic value of small morphological features.

Unfortunately, taxonomy and systematics generally are not popular in biology today, but correctly identifying organisms is at least as important now as ever before. In dinoflagellate research, the lack of a sound classification system causes uncertainties when trying to compare experimental results between different cultures supposedly of the same species (Dale, 1977a, pp. 249–251) and is a major reason why as yet there is little usable ecological data for the group (see discussion in section 4.5 under Cysts as indicators of dinoflagellate distribution). The extent of the problem is highlighted by the taxonomic chaos in toxic dinoflagellate studies. This is currently the most active field of biological research on dinoflagellates, but recent attempts by Loeblich and Loeblich (1979) and Taylor (1979) to change the names of common toxic species have done little to change the basic identification problem (see Taylor, 1975). Ecological and physiological research in this field is now proceeding in a taxonomic "vacuum" and presumably will have to be reevaluated when we understand basic information such as how many different organisms are involved and which are toxic.

For the phycologist, cysts are one of the newer taxonomic features that can be used to help strengthen dinoflagellate classification, but only within certain limits. For whereas all known living dinoflagellates have a motile stage (or its equivalent in certain specialized forms), most of them seem not to have cysts. The motile stage will therefore retain its taxonomic importance as the main stage on which overall classification of the group will be based. However, when cysts are present, they may offer a clearer indication of taxonomic subdivisions. Direct comparison of motile stage details for some peridinioid dinoflagellates reported by Balech (1974) with their cysts showed that cyst differences between species were more obvious; but cyst differences generally complemented Balech's taxonomic subdivisions based on relatively minor thecal details, suggesting that the two classification systems are generally compatible (Dale, 1977a, 1978).

The paleontological classification of dinoflagellates

In some respects fossil cysts are taxonomically easier to deal with than motile stages. The morphological range of body shape is about the same in both, if the whole known fossil record is considered, but the finer details on many cysts are probably not so complex as those of motile stages. In particular, paratabulation is seldom as detailed as tabulation, though some remarkable examples are being discovered in the fossil record. Recently, Piel and Evitt (1980) used extensive scanning electron microscopy to reveal extremely detailed paratabulation on cysts previously considered devoid of such, while Eaton (1980) showed a series of inferred major plate patterns for fossil dinoflagellates that are more complete than current information for many thecate dinoflagellates. But like the "neotaxonomist," the "paleotaxonomist" faces problems with many inadequate published descriptions of taxa, and with preservation. Cysts are basically resistant, but fossilization often results in flattening, breakage, folding, and sometimes oxidation, which obscures details.

Some of the most important differences between the biological and paleontological systems of classification concern available resources and basic approaches. There are many more "active" dinoflagellate taxonomists (i.e., those describing and classifying taxa) in paleontology than in biology, reflecting basic differences between the two fields. Micropaleontology is still in a very active exploration phase, with by far the greatest effort aimed at finding out which microfossils are present in a given geological time period. Much of this work is done by or for the oil industry and is usually applied in stratigraphic correlation problems. Even many industrial palynologists are thus actively engaged in the identification and description of new types of cysts. By comparison, the exploration phase in living dinoflagellates, seldom more than just a part of general phytoplankton investigations, represents a comparatively minor effort, and most biologists working with dinoflagellates do so with a minimum of taxonomic involvement.

The stratigraphic palynologist and the biologist differ significantly in their approaches to dinoflagellate classification. The palynologist is trying to break down the evolutionary record into ever smaller time units. The emphasis is therefore primarily on *differences* between forms, and the palynologist is not afraid to emphasize these differences in classification. In principle it is better to clearly separate even small morphological differences within the classification system; separate taxa may be combined later if warranted. The opposite philosophy often prevails in biology, where emphasis is on taxonomic continuity. Small morphological differences are often accommodated within one taxon, because they may represent natural variation, or the original description was not precise. There appears to be a working principle in biological literature that ultimate retrieval of ecological information is hampered by "splitting" and a change of name.

Although occasional fossil dinoflagellates closely resemble motile stages and have been sometimes regarded (and classified) as such, it is now generally accepted that known fossils are cysts. Therefore, the paleontological classification system is based on a different stage in the life cycle than that on which the biological system is based (cyst versus motile stages). Classification of the fossils was begun long before their dinoflagellate affinities were realized, and the two systems thus evolved separately. As with living dinoflagellates, the fossils were earlier classified as animals but are now generally classified as plants. As fossils, and representing one part of only some dinoflagellates, they are classified as form genera under the International Code of Botanical Nomenclature.

Paleontologists have good reasons for maintaining their own classification. They basically regard the two systems as separate but equal, with taxonomy of living dinoflagellates based on morphology of motile stages no more inherently "natural" than that of fossils based on cyst morphology (e.g., Evitt, 1970). The fossil taxon is based on a preserved type specimen, a fundamental difference from many living dinoflagellates that are very difficult to preserve. Throughout much of the fossil record, there is no possibility of direct reference to a holomorph including motile stage, and virtually no taxonomic overlap between the two classifications. Furthermore, a subjective morphological classification of cysts using binomial nomenclature based on the form genus is practical and has proved adequate for the needs of stratigraphic palynologists working with pre-Quaternary fossils. For this type of work, no great need was felt for a suprageneric classification, and genera often have been simply listed alphabetically in publications. The major paleontological failing therefore has been the obvious loss of phylogenetic information in a system not relating the various form genera to each other.

Currently, fossil dinoflagellate systematics is in an active phase of "consolidation," with major reassessment of inadequately described taxa (e.g., Stover & Evitt, 1978), and attempts to include phylogenetic information through lineages (e.g., Wall & Dale, 1968a; Williams, 1977) or to construct suprageneric classifications (e.g., Sarjeant & Downie, 1974; Norris, 1978). Results so far are encouraging. As better-documented stratigraphy and cyst morphology become available, the inevitable comparison is being made between this and information for living cysts in the search for phylogeny. In this way the important merging of biological and paleontological information is taking place without formal attempts to merge the two classifications.

Living cysts and classification

The study of living cysts offers the only way to directly link the biological and paleontological classifications of dinoflagellates. If fossil cysts were to be used only in applied stratigraphy, and if the interest in living dinoflagellates were restricted only to the plankton, there would be little incentive to explore such links. The broader interests of evolution, phylogeny, ecology, and paleoecology, however, require linkage between biology and paleontology, and for dinoflagellates the only "common denominator" is living cysts.

A necessary first step toward incorporating cysts into the biological classification is determining which living cysts correspond to which motile stages; it also offers an "independent check" on both classifications. Both are based on morphology, the fossils out of necessity because no other evidence is available, the living forms because the taxonomic value of other factors, such as biochemistry, is not yet understood (e.g., Ragan & Chapman, 1978, pp. 224–225). Biologists and paleontologists have classified motile stages and cysts, respectively, based on arbitrary evaluation of morphological criteria (i.e., without the possibility for genetic control). Obviously, if significant genetic morphological differences are expressed equally by cysts and motile stages, and if biologists and paleontologists are "correctly" (or even equally) interpreting the criteria reflecting these differences, then genera and species in the biological system should correlate with form genera and "cyst species" in the paleontological system.

They do not correlate; at least not most of those for which correlations are known, hence the lively discussion mentioned earlier in this section. The living fossil cysts that Wall and Dale (1966, 1968a, 1968b, 1969, 1970, 1971) correlated with motile stages included several examples in which one biological taxon correlated with one paleontological taxon (e.g., cysts of *P. bahamense* or *P. vancampoae*). The others were cysts of *Gonyaulax* species that correlated with many different paleontological form genera. Within these *Gonyaulax* species, again, some correlated with one paleontological taxon (e.g., *G. polyedra* or *G. grindleyi*), but many paleontological taxa (including at least five different form genera) correlated with just one closely related group of species, the *G. spinifera* group (see Fig. 45).

Results from correlating "fossil cysts" with motile stages may be variously interpreted depending on the point of view. Paleontologists cite the fact that eight cyst-based form genera probably correlate with the genus *Gonyaulax* as evidence of the inadequacies of the biological classification, and presume that *Gonyaulax* therefore is divisible into eight genera based on cysts (e.g., Reid, 1974; Reid & Harland, 1977). No mention is made in this view of inadequacies in the paleontological classification. On the other hand, biologists tend to suspect environmental (phenotypic) control of cyst morphology, especially since many of these different cyst morphotypes show differing patterns of distribution (Reid & Harland, 1977, p. 151). Both these views are probably oversimplifications.

As yet we do not know how to interpret these apparent correlations, but I suggest the question remains open until the following points are answered:

1. Are there relatively small but distinctive features on the motile stages that have not yet been identified? Certain orthoperidiniums were considered conspecific until smaller thecal features were reexamined, and these pointed to generic subdivisions (Balech, 1974; Dale, 1977a). Direct comparison of the biological and paleontological classification systems has suggested that larger cyst features are probably taxonomically "equivalent" to smaller thecal features (Dale, 1978). Kofoid (1911) subdivided the genus *Gonyaulax* into four subgenera and redescribed many species from a detailed study of plankton from the San Diego

region. Cysts still have not been correlated with most of these, and motile stages of species of *Gonyaulax* have not been examined in such detail from other regions (including those where most of the cysts in question are found). Wall (1975, Fig. 2) differentiates motile stages from the form genera *Operculodinium, Lingulodinium,* and *Spiniferites* and others based on apical thecal plates.

2. Palynologists usually presume that cyst morphology is genetically rather than environmentally controlled, but this view is based on very few examples of cyst morphologies being repeated through several generations (e.g., Wall & Dale, 1970, p. 54); most of the cysts in question have never been tested in this way. Therefore, the possibility remains that some cyst species could be phenotypic variants.

3. Is cyst morphology particularly less conservative than motile stage morphology in gonyaulacacean dinoflagellates? Although *Gonyaulax* is often used as an example of problems between the two systems, it is extreme rather than typical in this respect compared to all known living cysts.

4. Do the morphological differences between the various cysts in Fig. 45 warrant taxonomic separation at the genus level? Dale (1978) concluded that the paleontological classification of cysts was "overclassified" in the sense of overemphasizing small differences, especially in separating form genera. Some of the cysts discussed here may fall into this category. It is time to reexamine the living cysts using a more "biological approach" (i.e., emphasizing similarities more and differences less) and to thus reevaluate taxonomic criteria inherited from the paleontological system. For example, are the criteria used to separate form genera in fossil cysts acceptable as generic criteria when integrating cysts into the biological classification?

If there is one cyst feature typical for *Gonyaulax*, it is the precingular archeopyle. In most cases this is formed by loss of the third precingular paraplate (3‴), but in one undescribed species of the form genus *Spiniferites* (Stover & Evitt, 1978, p. 269) and a related form (Dale, unpublished, Fig. 36) and the monospecific form genus *Bitectatodinium* (Figs. 37, 45), two adjacent paraplates are lost (3‴ and 4‴?), whereas *G. polyedra* cysts lose three (sometimes five) adjacent paraplates. In view of the obvious tendency toward occasional loss of adjacent paraplates in archeopyle formation in gonyaulacacean cysts, I think this should not be used to separate otherwise identical cysts at the generic level. Wall and Dale (in Wall et al., 1973) emended the form genus *Lingulodinium* to include such archeopyle variations, and for the same reasons *Tectatodinium* and *Bitectatodinium* (Fig. 45) should not be separate form genera, and a new form genus should not be created for the species of *Spiniferites* with a 2p archeopyle previously mentioned. Similarly, it is worth considering afresh what levels of differences in ornament warrant generic separation. For example, when paratabulation and archeopyle are identical, as in the form genera *Impagidinium, Spiniferites,* and *Nematosphaeropsis* (Fig. 45), is generic separation warranted when the septa outlining paraplates in *Impagidinium* are "extended" out into processes as in *Spiniferites* (especially since "transition" forms are known, such as *Spiniferites inequalis;* Wall et Dale, Wall & Dale, 1973) or when these processes are joined by "septa" linking their outer tips (*Nematosphaeropsis*) instead of their bases (*Spiniferites*)? Even in the case of the morphologically dissimilar *Tectatodinium* and *Spiniferites*, we have a third line of morphological evidence that tends to support their closer relationship (as do motile stages) rather than more distant relationship (e.g., separate cyst "genera"). Species of both *Tectatodinium* (*T. psilatum*, Wall & Dale, 1973) and *Spiniferites* (*S. cruciformis*, Wall & Dale, 1973) probably "responded" morphologically in exactly the same way (by developing a characteristic cruciform body shape) in response to low-salinity environments in the Late Quaternary of the Black Sea (Wall et al., 1973).

Correlating living representatives of cysts previously classified as fossils with their motile stages thus far leaves us in a tantalizing state of knowledge. On the one hand, largely due to their commercial value in the oil industry, fossil cysts are being studied on a scale seldom before seen in paleontology. Much of this work is descriptive and must contain a wealth of phylogenetic and paleoecological information. On the other hand, extracting this information necessitates direct comparison with living dinoflagellates, and this can only take place through their cysts. Systematics is the "language" through which we communicate this type of descriptive science, but for dinoflagellates we have two separate systems (languages). Much of the communication problem between biology and paleontology of dinoflagellates thus focuses on resolving differences between these two languages as applied to living cysts. Results so far have shown us the extent of the language problem and indicated where we should seek at least some of the answers (e.g., points 1–4 above). As Reid and Harland (1977, p. 151) point out, we know enough of cyst/motile stage relationships to initiate a unified system of nomenclature for dinoflagellates. Understandably, this falls outside the main interests of industrial palynologists and biologists at present. With few palynologists having access to plankton culturing facilities (even if they were interested) and with as yet little or no input from biologists, progress toward integrating cysts into dinoflagellate classification is bound to be slow.

In the meantime, paleontologists studying cysts in Recent sediments (see references in Reid & Harland, 1977; Wall et al., 1977) or biologists studying living cysts face special nomenclatural problems. Presumably all these cysts have living motile stages, many of which will already be classified within the biological system. But the cysts fall into the following categories:

1. *Cysts correlated with their motile stages but not previously recorded as fossils* (e.g., cysts of *P. schwartzii*).

2. *Cysts correlated with motile stages but previously classified as fossils* (e.g., the fossil cyst species *Hemicystodinium zoharyii* (Rossignol) Wall, which on incubation proved to be the cyst of *P. bahamense*) – as discussed earlier, several different cysts may correlate with one motile stage according to present information.

3. *Cysts with established fossil-based names and not yet correlated with their motile stages* (e.g., living representatives of the form genus *Impagidinium*).

4. *As yet unidentified, undescribed cysts, morphologies of which may or may not suggest affinities with the paleontological or biological systems.*

Living cysts and cysts in Recent sediments offer maximum overlap of paleontological and biological information, and where possible this should be reflected in their classification. Both existing nomenclatural systems are inadequate for these cysts, most of which fall into category 4, but a unified system for these may still be a long time evolving, depending on the extent to which paleontologists and biologists become actively engaged in the problem. The best possible compromise must be sought in the meantime, and both a "paleontological" and a more biological approach have been suggested during the past few years. Both approaches are admittedly compromises until living cysts can be integrated into a unified classification system.

The paleontological approach to Recent and living cyst nomenclature is essentially an extension of the palynological system discussed earlier in this section. Reid (1974) outlined fully why paleontologists insist on maintaining a separate classification even for living cysts by creating new cyst-based taxa where necessary. This results in many new, narrowly defined form genera with only one or very few cyst species in each. Where possible these are related to their motile stages by including references to cyst/motile stage correlations, or by assembling cyst-based taxa into dinoflagellate families (e.g., gonyaulacacean or peridiniacean).

Dale (1978, and here) argues for a more "biological" approach. The main objection to the paleontological approach to living cysts is that it leads to a large increase in systematic literature without appreciable increase in either biological or paleontological knowledge. An alternative compromise is suggested, therefore, aimed at carrying maximum biological and paleontological information now toward gradual unification of the two systems. This begins by recognizing that two systems are in use now, and that in some cases cyst morphology currently offers a better subdivision of taxa than does known motile stage morphology. But particularly where cysts are already linked to a usable motile stage-based taxon (cyst category 1) or as yet unidentified (cyst category 4), I believe that new cyst-based nomenclature should not be created just to "artificially" maintain dual classification.

Bradford (1978, p. 195) presents the opposite point of view, justifying this as necessary for a "usable" classification system "to keep all cyst taxonomy uniform." For example, Bradford (1975) created new monospecific form genera, *Multispinula quanta* Bradford and *Omanodinium alticinctum* Bradford for the cysts of *P. conicum* and *Protoperidinium subinerme* (Paulsen) Loeblich. I consider these new names no more usable than the motile stage-based names. Furthermore, the new names carry neither paleontological (since these types are not previously described fossils) nor biological information (as do the motile stage-based names) and the ideal of "keeping cyst taxonomy uniform" thus seems poor justification for increasing the systematic literature to this extent. In order to communicate the known biological information for these (and similar) examples, Bradford has to note the correlation between his new taxa and their biological equivalents. In effect, for such forms this is "artificially" creating a new dual system out of one that is already unified.

Some motile stage/cyst holomorphs are recognizable already (cyst category 1), and I think they should be treated as such in an informal compromise approach to living cyst nomenclature. The other cysts could be treated as follows: Category 2 cysts would be named so as to reflect maximum biological and/or paleontological information; for example, the motile stage-based name should be used, if adequate (informal priority should be given to this rather than the paleontological name for living cysts), but the paleontological links should be noted. Paleontological names should be used where they obviously provide a better taxonomic separation (e.g., various cysts in the spinifera group of *Gonyaulax*), though a reevaluation of these may be required as discussed above. Category 3 cysts should carry their fossil names for now, though again reevaluation of these possibly resulting in taxonomic "consolidation" would be a useful stage toward eventual unification. Category 4 cysts require no names in an informal system as outlined here; numbers would suffice, but if names have to be applied, they should reflect likely biological and/or paleontological affinities where possible (e.g., *Spiniferites elongatus* Reid, 1974, reflects a useful paleontological link).

4.5 Cysts and dinoflagellate ecology

As with other organisms, the study of ecology of dinoflagellates is broadly concerned with finding out which species live where, and how and why they do so. Ecological studies of planktonic motile stages of dinoflagellates have not progressed very far. For reasons discussed below, plankton studies are still trying to establish which species live where, and questions concerning how and why they do so have as yet received little attention. Even basic elements of the dinoflagellate "swimming strategy" in contrast to the "sinking strategy" of diatoms, or the possible combination of both found in coccolithophorids remain largely undefined.

Cyst studies mainly by paleontologists are revealing a benthic view of dinoflagellate ecology that offers the biologist new insight both into which species live where and how and why at least some of them do so.

Cysts as indicators of dinoflagellate distribution

Paleontologists have begun to study the distribution of living cysts as an important step toward understanding how distribution of fossil cysts may be used to interpret paleoecology. Results in the past few years have clearly demonstrated that cyst distribution expresses enough ecological information for the fossils to be useful indicators of paleoclimate and paleosalinity. This paleontological approach to dinoflagellate ecology is of potential value to the biologist. On first consideration, it may seem unreasonable to seek ecological information from cysts probably representing only a selective group of dinoflagellates, but this has to be judged in light of the special difficulties limiting studies of the distribution of planktonic motile stages.

Problems involved in studying distribution of planktonic dinoflagellates. Two main problems hinder distribution studies of planktonic dinoflagellates:

1. *Sample coverage.* Dinoflagellate populations have only been studied in adequate detail from a few regions (e.g., northern North Altantic, Lebour, 1925; British Isles, Dodge & Hart-Jones, 1978; West Florida, Steidinger & Williams, 1970; Indian Ocean, Taylor, 1976, Southwest Atlantic, Balech, 1979). As yet, there is little overlap between these regions, and biogeographical boundaries for most species remain unknown. Otherwise, general phytoplankton counts from other regions offer usable information for the more easily identified dinoflagellates (e.g. *Ceratium* species, Graham & Bronikovsky, 1944), but questionable information for many others (discussed in 2). In neritic environments, strongly influenced by a seasonally changing species succession, or in some open ocean environments with different species at various depths, it is doubtful that enough plankton samples will be examined in enough detail to define biogeographical boundaries for dinoflagellates.

2. *Identification problems.* Given the difficulties with inadequate descriptions and the increasing realization that comparatively small morphological features probably are taxonomically important, correctly identifying many dinoflagellate motile stages often is difficult and very time-consuming for the taxonomic specialist. For the general phytoplankton ecologist counting samples, it is practically impossible. The commonly used inverted microscope technique for counting plankton does not allow close scrutiny of small thecal plates, and there are practi-

cal limits to the time which can be taken to identify each specimen. In practice, this results in plankton lists that include some dinoflagellates identified from major easily recognizable features. Other dinoflagellates, however, are of more questionable identity; these are identified to generic level only, or included under the most appropriate species name available in the literature even though some uncertainty remains (e.g., the original description may be too imprecise, the original description may contain fine details which cannot be confirmed, or small discrepancies between the published description and an observed specimen may be considered possible natural variation). Balech (1970) showed that misidentifications significantly distorted the previous view of Antarctic plankton.

These seemingly unsurmountable problems with sampling and identification will probably severely limit our ability to map the biogeographical distribution of many planktonic dinoflagellates in the forseeable future.

Paleontological studies of cyst distribution. The potential use of fossil cysts as paleoecological indicators is of interest to both pre-Quaternary and Quaternary geology. The pre-Quaternary palynologist trying to use changes in fossil cyst assemblages for biostratigraphy is interested in differentiating between changes caused by paleoecology and those caused by evolution. This is particularly important when correlating between different sedimentary basins that might have originally represented very different environments (e.g., different paleolatitudes during the same time. In addition, paleogeographical reconstructions are sometimes an important part of oil exploration, and relevant information from cysts would be particularly valuable since often these are already used for biostratigraphy. The Quaternary palynologist is dealing with too short a period for much usable evolutionary change and instead builds up a stratigraphy based on the paleoecological changes caused by dramatic oscillations in paleoclimate. Pollen analysis has long been used for nonmarine sediments, but as early as 1954 Erdtman suggested the potential use of fossil cysts as paleoclimatic indicators in marine sediments. This potential is only now beginning to be realized.

Several palynologists have studied present-day distribution of cysts in Recent sediments as a means toward better understanding fossil cyst distributions in paleontology (e.g., Rossignol, 1964; Davey, 1971; Williams, 1971; Reid, 1972, 1975; Davey & Rogers, 1975; Dale, 1976; Harland, 1977; Wall et al., 1977). These studies mostly employed standard palynological preparation techniques which probably destroy some cysts (Dale, 1976) but give results that are more comparable with fossil assemblages. The sediments used probably vary somewhat in age, the deep-ocean sediments in particular probably including cysts that are many thousands of years old. However, the few examples allowing comparison with living cysts suggest that main distribution patterns in such palynological preparations, at least from shallower water, probably reflect living cyst patterns (Dale, 1976; Wall et al., 1977). Recent studies using deep sea sediment traps, however, suggest that comparable cyst distribution patterns in some deep-ocean sediments clearly do not reflect living cyst distribution (Dale, in press.)

Results from Wall et al. (1977) are particularly pertinent to the discussion here. This work produced an ecological classification for many cysts, based on a quantitative analysis of 168 cyst assemblages in modern marine sediments largely from the North and South Atlantic oceans. Two main trends identified in the cyst distribution data were variations from inshore to offshore and with latitude. Cyst variation involved associations of species and species diversity, but of more interest here is the distribution patterns of individual taxa. Some common types were found to be cosmopolitan, but others showed clearly restricted distributions. Wall et al. (1977, pp. 175–176) were able to classify the cysts into four main groups according to their inshore to offshore distributions (estuarine, estuarine–neritic, neritic, and oceanic), and into nine categories within these groups according to their latitudinal distribution (e.g., temperate estuarine species). Overall cyst distribution was obviously related to water mass characteristics, as suggested also by Williams (1971), Reid (1972), and Reid and Harland (1977). Most significantly for this discussion, it was possible to relate the cyst biogeographical data thus produced to current ecological theories (Wall et al., 1977, pp. 179–190), a stage not yet reached for planktonic dinoflagellates.

Living cysts as indictors of dinoflagellate distribution. Studying cysts in bottom sediments offers the biologist a new approach to dinoflagellate ecology. This benthic view is admittedly restricted, since many species probably do not form cysts. For cyst-forming species, however, this approach at least offers the possibility of overcoming the main difficulties presently limiting ecological studies of motile stages in plankton. Sampling difficulties are minimal, with a single sediment sample providing information that is integrated from the whole water column and for at least one season (if living cysts are studied), or up to many years (if fossilizable cysts are studied using palynological methods). Identifying taxa is also less problematic, with cyst morphology generally less complex and less conservative than that of corresponding motile stages.

There are still many unanswered questions concerning cyst distribution patterns. Does cyst distribution in bottom sediments accurately reflect the planktonic distribution of a species? Are distribution patterns revealed by cysts also typical for non–cyst-forming species, or do the cysts represent a significantly different strategy producing different patterns?

While not answering such questions completely, information to date suggests some intriguing possibilities. In a study to be published elsewhere, I recently compared known present-day cyst distribution with standard biogeographical zones (arctic, cold temperate, warm temperate, tropical) based on distribution of other marine organisms such as mollusks. Though cyst data are as yet far less comprehensive, they suggest biogeographical zones comparable with those established for the other organisms (see Fig. 46). Furthermore, consistent changes in fossil cysts from Quaternary sediments suggest migration of these cyst zones in response to climatic change. Thus, cysts appear to give "plausible" ecological information. If some cysts are restricted to standard biogeographic zones, however, does this imply a definitive benthic role in dinoflagellate ecology? Temperature is a likely major factor controlling these biogeographical zones, but as yet we do not know the extent to which latitudinal distribution of some species is limited by temperature effects on the cyst or the motile stage.

Probable functions of cysts

Wall (1971) fully discussed the possible functions of cysts. This, together with subsequent information of life cycles and cyst distribution, suggests ways in which cysts may affect how and why some species live where they do. The probable main functions of cysts (discussed previously here) may be summarized:

1. *Sexual cycles.* Accumulating evidence strongly suggests that cysts are zygotes whose main function is probably nuclear replenishment through meiosis. It is not yet clear how widespread sexual cycles are in dinoflagellates, or how important for populations under favorable growth conditions. However, cyst formation under seemingly favorable conditions in the tropics and under culture conditions where vegetative cells continued to multiply suggest this is not just a response to unfavorable conditions.

2. *Protection.* The cyst apparently serves a protective role, allowing the individual to withstand conditions that would be adverse for the motile cell. For example, cysts in temperate regions routinely overwinter in lower temperatures than the motile cells tolerate and may survive at low temperatures for many years if not returned to conditions favorable for the motile stages.

3. *Propagation.* Cysts probably act as seed populations, a resting period and dormancy ensuring that a new population of motile stages can be established during favorable intervals in a fluctuating environment. This probably allows some species to extend their ranges beyond the climatic boundaries in which motiles alone could survive year round.

4. *Dispersion.* Available evidence suggests that many cysts sink to the bottom, but as fine "silt particles" they presumably are sometimes transported by currents and dispersed within the sedimentary regime. Cysts may thus help to extend a species range either through "invasion" of new territory by dispersed cysts that later establish a motile population or through occasional "invasion" by motile stages that form cysts and a new seed population. The northeast coast of the United States may offer a striking example of this – southern Massachusetts has had repeated outbreaks of toxic red tide each year since the first ever recorded there in 1972. In this case, cysts probably turned what otherwise might only have been a temporary invasion by *G. excavata* in 1972 into a more permanent invasion (Dale, 1977b).

Functional morphology in cysts?

Williams (1977) summarized the attempts by paleontologists to use cysts as paleoenvironmental indicators. Inevitably this has included speculation concerning functional morphology. For example, Davey (1970) suggested that domination by chorate cysts (i.e., those with outgrowths such as processes) in Cretaceous warmer-water assemblages reflected the need for greater flotation in warmer and therefore less dense water. Vozzhennikova (1965) suggested that thick-walled fossil cysts were concentrated in neritic zones and thin-walled cysts with added flotation from processes were more typically open marine.

Studies of Recent cyst distribution cited earlier in this section suggest no such general trends in functional morphology. Recent cyst assemblages in marine environments often largely represent species of two genera, *Gonyaulax* and *Protoperidinium*; the few known freshwater cysts mostly belong to *Peridinium*. Almost all gonyaulacacean cysts have moderate to long processes, whereas cysts of *Protoperidinium* species are mostly smooth walled and lacking extensive ornamentation. Recent cyst assemblages from tropical to cold-temperate regions are often dominated by one of several cosmopolitan species, mostly gonyaulacacean cysts with well-developed processes; *Protoperidinium* cysts are usually present but not dominant. There is some evidence from Quaternary glacial sediments and Recent equivalents suggesting that smooth-walled *Protoperidinium* cysts (e.g., *P. conicoides* [Paulsen] Balech) may dominate near permanent or semipermanent sea ice. However, this more likely reflects the basic independence from light of the largely nonphotosynthetic *Protoperidinium* species rather than their cyst morphology (Dale, unpublished).

Open-ocean cysts living now do not have thinner walls, and they show no general tendency toward increased flotation. At least one outer neritic/oceanic cyst, *Nematosphaeropsis labyrinthus* (Ostenfeld) Reid

(Figs. 34, 35), could be interpreted as showing adaptation toward reduced sinking rate (with long processes linked together at their outer tips by thin strands), as could the oceanic cyst type, *Impagidinium aculeatum* (Wall) Stover et Evitt (with septa drawn out into extensive ridges). Examination of the material from deep-sea sediment traps, however, has shown that organic cysts also include many not obviously adapted to increased flotation, and by far the most typical open-ocean cysts are thick-walled calcareous cysts (e.g., Figs. 29–31) seemingly adapted to a sinking strategy (Dale, in press).

If extensions (such as processes) from the cyst wall are adaptations toward reduced sinking, then evidence so far suggests the possibility of both sinking and floating strategies in neritic and open-ocean cysts.

Cysts: a survival strategy?

It may seem premature to address the question of whether or not cysts are a survival strategy, since we know so little about the basic strategies of motile dinoflagellates. However, the cyst information summarized here at least allows us to consider this possibility.

That cyst-producing dinoflagellates have survived for long periods of geological time is well documented from the fossil record. The oldest, generally accepted fossil cyst is *Arpylorus antiguus* Calandra from the Upper Silurian (recently reexamined by Sarjeant, 1978). The traditional use of the archeopyle to differentiate cysts from acritarchs, however, gives a purely artificial separation (discussed in 4.2 under Cyst morphology), and some older Lower Paleozoic acritarchs may well represent cysts. We are unable to assess any possible selective advantage that cyst formation may have had over non–cyst-formers, since only the cysts of dinoflagellates fossilized. However, it seems likely that some dinoflagellates have survived long periods without cysts. For example, *Ceratium*-like cysts are known from marine Cretaceous sediments (Wall & Evitt, 1975), but not from more recent deposits. Marine species of *Ceratium* living now seem not to form cysts and therefore may be the product of at least 70 million years of evolution without cysts. Such gaps in the record, where fossilizable cysts were not produced, severely limit the paleontologist's view of dinoflagellate evolution (Dale, 1976; Piel & Evitt, 1980).

The term *survival strategy* is used here for apparent adaptations by the individual that allow it to continue to exist. Cysts thus may be interpreted as a survival strategy allowing some individuals to survive under conditions that are demonstrably adverse for the motile stage. For example, in the laboratory, some cultures left to age without transfer to fresh medium may eventually contain only cysts on the bottom, the medium no longer allowing survival of the motile stages. However, it is virtually impossible to generalize from the individual to a natural population. Even in the most extreme overwintering situation, it is impossible to know if a few motile stages survive the winter at undetectable concentrations. Alternatively, the motile population may maintain contact with more favorable waters (e.g., the coastal current and associated offshore waters off the Norwegian coast), reinvading nearshore waters as they warm up seasonally.

From the evidence presented here, it could be argued that cysts represent a basic strategy whereby planktonic organisms produce benthic resting stages that may offer the individual survival, at least under extreme conditions. Furthermore, evidence is accumulating suggesting this is a basic strategy found in a wide range of other planktonic organisms. In addition to those dealt with elsewhere in this volume, we now know that some copepods produce "overwintering" eggs (Grice & Gibson, 1975, 1977), tintinnids produce "cysts" (Reid & John, 1978), and several possible cysts from benthic sediments that in fact produced ciliates following incubation (Dale, unpublished).

In such cases, where a group of organisms traditionally has been first studied as planktonic organisms, it is perhaps to be expected that subsequently found resistant benthic stages will be considered a survival strategy for the plankton. In some respects, this situation is opposite to that of some mollusks, for example, traditionally studied as benthic organisms and subsequently found to produce planktonic larvae. These, however, are probably not simply planktonic organisms "surviving" by producing benthic resting stages, or benthic organisms "surviving" by producing planktonic larvae. More likely, these organisms have evolved an alternation of generations, one planktonic and one benthic, the combination of which offers *one* viable "life strategy." Considered from this perspective, many of the "survival" aspects of cysts may be adaptations toward survival of the cysts in the particularly rigorous benthic environment.

Irrespective of the extent to which cysts represent a survival strategy, cyst formation is clearly only one alternative, since many species apparently survive just as well without producing cysts. Some clues to other alternative strategies are probably to be found by studying the life cycles of non–cyst-forming dinoflagellates in temperate freshwater lakes.

4.6 Application of cyst work to living dinoflagellate studies

Cysts in bottom sediments may be of practical use to the biologist. Attention is particularly drawn to their use in the following important aspects of dinoflagellate research.

1. *Culturing dinoflagellates.* Wall et al. (1967) described techniques for starting cultures from cysts. This is usually much easier than isolating motile cells from plankton and offers a higher success rate for motile

stages excysted directly into media rather than undergoing media shock on transfer from natural waters. Cysts may be stored easily in a refrigerator for many years as "potential cultures"; an alternative to maintaining cultures of motile cells that are used only occasionally.

2. *Basic surveys.* The integrated information from a sediment sample is a useful addition to plankton surveys, especially where plankton sampling is limited (e.g., expeditions to remote areas). In markedly seasonal environments, a combination of plankton samples (showing the motile stages at a given time) and sediment samples (possibly showing other species encysted) is particularly useful. Even in combination with extensive plankton surveys, cysts may suggest the undetected presence of species that are more difficult to identify as motile stages (e.g., Dale, 1976, added several species to the plankton list for Trondheimsfjord).

3. *Red tides and toxic dinoflagellates.* Lewis et al. (1979) showed the possibility of mapping cysts of the toxic *G. excavata.* This may reveal areas of higher or lower toxicity that may help in siting toxicity monitoring stations or shellfish-culturing projects. A further application of cyst work is identifying causative organisms by culturing them from cysts remaining in sediments after toxic blooms. By the time it is realized that shellfish are poisoning humans following such blooms, the causative dinoflagellate often is no longer present in the plankton to be identified (Dale, 1979).

References

Anderson, D. M. (1980). Effects of temperature conditioning on development and germination of *Gonyaulax tamarensis* (Dinophyceae) hypnozygotes. *Journal of Phycology*, **16**, 166–172.

Anderson, D. M., & Morel, F. (1979). The seeding of two red tide blooms by the germination of benthic *Gonyaulax tamarensis* hypnozygotes. *Estuarine and Coastal Marine Science,* **8**, 279–293.

Anderson, D. M., & Wall, D. (1978). Potential importance of benthic cysts of *Gonyaulax tamarensis* and *G. excavata* in initiating toxic dinoflagellate blooms. *Journal of Phycology,* **14**, 224–234.

Balech, E. (1970). The distribution and endemism of some Antarctic microplanktons. In *Antarctic Ecology*, ed. M. W. Holdgate, Vol. 1, pp. 143–147. New York: Academic Press.

– (1974). El género *"Protoperidinium"* Bergh, 1881 (*"Peridinium"* Ehrenberg; 1831, partim). *Revista del Museo Argentino de Ciencias Naturales "Bernardino Rivadavia" e Instituto Nacional de Investigación de las Ciencias Naturales, Hidrobiologia,* **4**, 1–79.

– (1979). Dinoflagelados de la campaña oceanográfica Argentina Islas Orcadas 0675. *Armada Argentina servicio de hidrografia naval Público,* **665**, 7–76.

Bibby, B. T., & Dodge, J. D. (1972). The encystment of a freshwater dinoflagellate: a light and electron microscopical study. *British Phycological Journal,* 7, 85–100.

Boltovskoy, A. (1973). Formación del arqueopilo en tecas de dinoflagelados. *Revista Española de Micropaleontologia,* **5**, 81–98.

Bourne, N. (1965). Paralytic shellfish poison in the sea scallops *Placopecten magellanicus,* Gemlin. *Journal of Fisheries Research Board of Canada,* **22,** 1137–1149.

Braarud, T. (1945). Morphological observations on marine dinoflagellate cultures. *Avhandling av det Norske Vitenskapelige Akademi, Oslo.* **1943–3,** 1–18.

Bradford, M. R. (1975). New dinoflagellate cyst genera from the recent sediments of the Persian Gulf. *Canadian Journal of Botany,* **53,** 3064–3074.

– (1978). Acritarchous cysts of *Peridinium faeroense* Paulsen: implications for dinoflagellate systematics. A discussion. *Palynology,* **2,** 195–197.

Brasier, M. D. (1980). *Microfossils,* pp. 1–193. London: Allen & Unwin.

Brooks, J., & Shaw, G. (1973). The role of sporopollenin in palynology. In *Problems of Palynology. Proceedings of 3rd International Palynology Conference,* ed. M. I. Neustadt, pp. 80–91.

Buchanan, R. J. (1968). Studies at Oyster Bay in Jamaica, West Indies. IV. Observations on the morphology and asexual cycle of *Pyrodinium bahamense* Plate. *Journal of Phycology,* 4(4), 272–277.

Cleve, P. T. (1900). The plankton of the North Sea, English Channel, and the Skagerak in 1898. *Svenska Vetenskapsakademien, Handlingar,* Bd. 32, No. 8.

Dale, B. (1976). Cyst formation, sedimentation, and preservation: factors affecting dinoflagellate assemblages in Recent sediments from Trondheimsfjord, Norway. *Review of Palaeobotany and Palynology,* **22,** 39–60.

– (1977a). New observations on *Peridinium faeroense* Paulsen (1905) and classification of small orthoperidinioid dinoflagellates. *British Phycological Journal,* **12,** 241–253.

– (1977b). Cysts of the toxic red-tide dinoflagellate *Gonyaulax excavata* (Braarud) Balech from Oslofjorden, Norway. *Sarsia,* **63,** 29–34.

– (1978). Acritarchous cysts of *Peridinium faeroense* Paulsen: implications for dinoflagellate systematics. *Palynology,* **2,** 187–193.

– (1979). Collection, preparation and identification of dinoflagellate resting cysts. In *Toxic Dinoflagellate Blooms, Proceedings of the Second International Conference on Toxic Dinoflagellate Blooms*, ed. D. L. Taylor & H. H. Seliger, pp. 443–452. New York: Elsevier North-Holland.

– in press. Dinoflagellate contributions to the open ocean sediment flux. *Micropaleontology.*

Dale, B., Yentsch, C. M., & Hurst, J. W. (1978). Toxicity in resting cysts of the red-tide dinoflagellate *Gonyaulax excavata* from deeper water coastal sediments. *Science,* **201** (4362), 1223–1225.

Davey, R. J. (1970). Non-calcareous microplankton from the Cenomanian of England, northern France and North America, II. *British Museum* (*Natural History*) *Bulletin of Geology,* **18,** 333–397.

– (1971). Palynology and palaeo-environmental studies, with special reference to the Continental Shelf sediments of South Africa. In *Proceedings of 2nd Planktonic Conference, Rome, 1970,* pp. 331–347.

Davey, R. J., & Rogers, J. (1975). Palynomorph distribution in Recent offshore sediments along two traverses off South West Africa. *Marine Geology,* **18,** 213–225.

Deflandre, G. (1933). Note Préliminaire sur les péridinien fossile *Lithoperidinium oamaruense* n.g., n. sp. *Zoological Society of France Bulletin,* **58,** 265–273.

– (1947). *Calciodinellum* nov. gen., premier répresentant d'une famille nouvelle de Dinoflagellés fossiles a theque calcaire. *Comptes Rendus Hebdomadaires des seances de l'Academie des Sciences* **224,** 1781–1782.

– (1948). Les Calciodinellidés – dinoflagellés fossiles a theque calcaire. *Botaniste*, 34, 191–219.

Dodge, J. D., & Hart-Jones, B. (1978). Further progress with mapping the distribution of marine dinoflagellates around the British Isles. *British Phycological Journal*, 13(2), 199.

Dürr, V. G. (1979). Electron microscope studies on the theca of dinoflagellates. 3. The cyst of *Peridinium cinctum*. *Archives für Protistenkunde*, 122, 121–139.

Eaton, G. L. (1980). Nomenclature and homology in peridinialean dinoflagellate plate patterns. *Paleontology*, 23(3), 667–688.

Ehrenberg, C. G. (1838). Über das Massenverhältnis der jetzt lebenden Kiesel-infusiorien und über ein neues Infusorien-Conglomerat als Polierschiefer von Jastraba in Ungarn. *Abhandlungen der Königlichen Akademie der Wissenschaften zu Berlin* (1836), 1, 109–135.

Eisenack, A. (1935). Mikrofossilien aus Dogger geschieben Ostpreussens. *Zeitschrift Geschiebeforschung*, 11, 167–184.

– (1936). *Eodinia pachytheca* n.g., n.sp. ein primitiver Dinoflagellat aus einem Kelloway-Geschiebe Ostpreussens. *Zeitschrift Geschiebeforschung*, 12, 72–75.

Elbrächter, M., & Drebes, G. (1978). Life cycles, phylogeny and taxonomy of *Dissodinium* and *Pryocystis* (Dinophyta). *Hegoländer Wissenschaftliche Meeresuntersuchungen*, 31, 347–379.

Erdtman, G. (1949). Palynological aspects of the pioneer phase in the immigration of the Swedish flora II. Identification of pollen grains in Late Glacial samples from Mr. Omberg, Ostrogothia. *Svensk Botanisk Tidskrift*, 43, 46–55.

– (1950). Fynd av *Hystrichosphaera furcata* i Gullmaren. *Geologiska Föreningen i Stockholmen Förhandlingar*, 72, 221.

– (1954). On pollengrains and Dinoflagellate cysts in the Firth of Gullmaren, S. W. Sweden. *Botaniska Notiser* 1954, 103–111.

Erėn, J. (1969). Cyst formation in *Peridinium cinctum* fa. *Westii*. *Journal of Protozoology*, 16, Suppl. 35.

Evitt, W. R. (1961). Observations on the morphology of fossil dinoflagellates. *Micropaleontology*, 7, 385–420.

– (1963). A discussion and proposals concerning fossil dinoflagellates, hystrichospheres, and acritarchs, I. *Proceedings of the National Academy of Sciences of the United States of America*, 49, 158–164.

– (1967a). Progress in the study of fossil *Gymnodinium* (Dinophyceae). *Review of Palaeobotany and Palynology*, 2, 355–363.

– (1967b). Dinoflagellate studies. II. The archeopyle. *Stanford University Publications, Geological Science*, 10(3), 1–88.

– (1969). Dinoflagellates and other organisms in palynological preparations. In *Aspects of Palynology*, ed. R. H. Tschudy & R. A. Scott, pp. 439–479. New York: Wiley–Interscience.

– (1970). Dinoflagellates – a selective review. *Geoscience and Man*, 1, 29–45.

Evitt, W. R., & Davidson, S. E. (1964). Dinoflagellate studies. 1. Dinoflagellate cysts and thecae. *Stanford University Publications, Geological Science*, 10(1), 1–12.

Evitt, W. R., Lentin, J. K., Millioud, M. E., Stover, L. E., & Williams, G. L. (1977). Dinoflagellate cyst terminology. *Canadian Geological Survey, Paper* 76–24, 1–11.

Fritsch, F. E. (1956). *The structure and reproduction of the algae*, Vol. 1. Cambridge: Cambridge University Press.

Gocht, H. (1979). Correlation of overlapping system and growth in fossil dinoflagellates (*Gonyaulax* group). *Neues Jahrbuch für Geologie und Paläontologie, Abhandlungen*, **157**(3), 344–364.

Gocht, H., & Netzel, H. (1976). Reliefstrukturen des Kreide-Dinoflagellaten *Palaeoperidinium pyrophorum* im Vergleich mit Panzer-Merkmalen rezenter *Peridinium*-Arten. *Neues Jahrbruch für Geologie und Paläeontologie, Abhandlungen*, **152**, 380–413.

Graham, H. W., & Bronikovsky, N. (1944). The genus *Ceratium* in the Pacific and North Atlantic Oceans. *Publications of the Carnegie Institution*, **565**(7), 1–209.

Grice, G. D., & Gibson, V. R. (1975). Occurrence, viability and significance of resting eggs of the Calanoid copepod *Labidocera aestiva*. *Marine Biology*, **31**, 335–337.

–(1977). Resting eggs in *Pontella meadi* (Copepoda: Calanoida). *Fisheries Research Board of Canada*, **34**, 410–412.

Harland, R. (1971). A summary review of the morphology and classification of the fossil Peridiniales (Dinoflagellates) with respect to their modern representatives. *Geophytology*, 1(2), 135–150.

– (1974). Quaternary organic-walled microplankton from boreholes 71/9 and 71/10. In *The Geology of the Sea of the Hebrides*, Report of the Institute of Geological Sciences No. 73/14, 1, ed. P. E. Binns, R. McQuillin, & N. Kenolty.

– (1977). Recent and Late Quaternary (Flandrian and Devensian) Dinoflagellate cysts from marine continental shelf sediments around the British Isles. *Palaeontographica, B*, **164**, 87–126.

Hensen, V. (1887). Über die Bestimmung des Plankton's – oder des im Meere treibenden Materials and Pflanzen und Tieren. *Berichte der Kommission für Wissenschaftliche Untersuchungen der deutschen Meere, Kiel*, **1882–1886**, 5.

Himes, M., & Beam, C. L. (1975). Genetic analysis in the dinoflagellate *Crypthecodinium* (*Gyrodinium*) *cohnii:* evidence for unusual meiosis. *Proceedings of the National Academy of Sciences of the United States of America*, **72**(11), 4546–4549.

Huber, G., & Nipkow, F. (1922). Experimentelle Utersuchungen über die Entwicklung von *Ceratium hirundinella* O.F.M. *Zeitschrift für Botanik*, **14**, 337–371.

– (1923). Experimentelle Untersuchungen über die Entwicklung und Formbildung von *Ceratium hirundinella* O. Fr. Mull. *Flora New Series*, **16**, 114–215.

Iversen, J. (1936). Sekundäres Pollen als Fehlerquelle. Eine Korrektionsmethode zur Pollenanalyse minerogener Sedimente. *Danmarks Geologiske Undersøgelse*, 2(15), 1–24.

Jux, U. (1976). Über den Feinbau der Wandungen bei *Operculodinium centrocarpum* (Deflandre & Cookson) Wall 1967 und *Bitectatodinium tepikiense* Wilson 1973. *Palaeontographica*, *B*, **155**, 149–156.

Kofoid, C. A. (1907). The plates of *Ceratium* with a note on the unity of the genus. *Zoologische Anzeiger*, **32**, 177–183.

– (1909). On *Peridinium steinii* Jörgensen, with a note on the nomenclature of the skeleton of the Peridinidae. *Archives für Protistenkunde*, **16**, 25–47.

– (1911). Dinoflagellates of the San Diego region. IV. The genus *Gonyaulax*, with notes on its skeletal morphology and a discussion of its generic and specific characters. *University of California Publications in Zoology*, **8**, 187–286.

Lebour, M.V. (1925). *The Dinoflagellates of Northern Seas.* pp. 1–250, Plymouth: Marine Biological Association of U.K.

Lefévre, M. (1933). Recherches sur des Péridiniens fossiles des Barbades. *Museum National d'Histoire Naturelle, Paris,* Bulletin, sér. 2, (5), 415–418.

Lentin, J. K., & Williams, G. L. (1975). Fossil dinoflagellates – index to genera and species. *Canadian Journal of Botany*, **53** (Suppl. 1), 2147–2157.

–(1976). A monograph of fossil peridinioid dinoflagellate cysts. *Bedford Institute of Oceanography Report* BI-R-75-16, pp. 1–237.

Lewis, C. M., Yentsch, C. M., & Dale, B. (1979). Distribution of *Gonyaulax excavata* resting cysts in sediments of the Gulf of Maine. In *Toxic Dinoflagellate Blooms, Proceedings of the Second International Conference on Toxic Dinoflagellate Blooms*, ed. D. L. Taylor & H. H. Seliger, pp. 235–238. New York: Elsevier North-Holland.

Lister, T. R. (1970). A monograph of the acritarchs and chitinozoa from the Wenlock and Ludlow Series of the Ludlow and Millichope areas, Shropshire. *Palaeontographical Society* (*Monograph*), **124**, 1–100.

Loeblich, A. R., & Loeblich, L. A. (1979). The systematics of *Gonyaulax* with special reference to the toxic species. In *Toxic Dinoflagellate Blooms, Proceedings of the Second International Conference on Toxic Dinoflagellate Blooms.* ed. D. L. Taylor & H. H. Seliger, pp. 41–46. New York: Elsevier North Holland.

Lohmann, H. (1904). Eier und sogenannte Cysten der Plankton-Expedition. *Ergebnisse der Plankton-Expedition der Humboldt-Stiftung*, new series, 4, 1–62.

–(1910). Eier und Cysten des nordischen Planktons. *Nordisches Plankton*, Zoologischer Teil 2, 1–20.

McKee, E. D., Chronic, J., & Leopold, E. B. (1959). Sedimentary belts in lagoon of Kapingamarangi atoll. *Bulletin of the American Association of Petroleum Geology*, **43**, 501–562.

Mantell, G. A. (1845). Notes of a microscopical examination of the chalk and flint of southeast England, with remarks on the Animalculities of certain Tertiary and modern deposits. *Annals and Magazine of Natural History*, **16**, 73–88.

Manum, S. B. (1978). Two new Tertiary dinocyst genera from the Norwegian Sea: *Lophocysta* and *Evittosphaerula. Review of Palaeobotany and Palynology*, **28**, 237–248.

Morey-Gaines, G., & Ruse, R. H. (1980). Encystment and reproduction of the predatory dinoflagellate *Polykrikos Kofoidi* Chalton (Gymnodiniales). *Phycologia*, **19**(3), 230–236.

Morzadec-Kerfourn, M. T. (1976). La signification écologique des dinoflagellés et leur intérêt pour l'étude des variations du niveau marin. *Revue de Micropaléontologie*, **18**(4), 229–235.

Nordli, E. (1951). Resting spores in *Gonyaulax polyedra* Stein. *Nytt Magasin for Naturvidenskapene*, **88**, 207–212.

Norris, G. (1978). Phylogeny and a revised supra-generic classification for Triassic-Quaternary organic-walled dinoflagellate cysts (Pyrrhophyta). 1. Cyst terminology and assessment of previous classifications. *Neues Jahrbuch für Geologie und Palaeontologie, Abhandlungen*, **155**, 300–317.

Norris, G., & McAndrews, J. H. (1970). Dinoflagellate cysts from postglacial lake muds, Minnesota (U.S.A.). *Review of Palaeobotany and Palynology*, **10**, 131–156.

Pfiester, L. A. (1975). Sexual reproduction of *Peridinium cinctum* f. *ovaplanum* (Dinophyceae). *Journal of Phycology*, **11**, 259–265.
– (1976). Sexual reproduction of *Peridinium willei* (Dinophyceae). *Journal of Phycology*, **12**, 234–238.
– (1977). Sexual reproduction of *Peridinium gatunense* (Dinophyceae). *Journal of Phycology*, **13**, 92–95.
Pfiester, L. A., & Skvarla, J. J. (1979). Heterothallism and thecal development in the sexual life history of *Peridinium volzii* (Dinophyceae). *Phycologia*, **18**, 13–18.
– (1980). Comparative ultrastructure of vegetative and sexual thecae of *Peridinium limbatum* and *Peridinium cinctum* (Dinophyceae). *American Journal of Botany*, **67**(6), 955–958.
Piel, K. M., & Evitt, W. R. (1980). Paratabulation in the Jurassic dinoflagellate genus *Nannoceratopsis* and a comparison with modern taxa. *Palynology*, **4**, 79–104.
Prakash, A. (1963). Source of paralytic shellfish toxin in the Bay of Fundy. *Journal of the Fisheries Research Board of Canada*, **20**, 983–996.
Prakash, A., Medcof, J. C., & Tennant, A. D. (1971). Paralytic shellfish poisoning in eastern Canada. *Fisheries Research Board of Canada Bulletin* 177.
Ragan, M. A., & Chapman, D. J. (1978). *A Biochemical Phylogeny of the Protists*. New York: Academic Press.
Reid, P. C. (1972). Dinoflagellate cyst distribution around the British Isles. *Journal of the Marine Biological Association of the United Kingdom*, **52**, 939–944.
– (1974). Gonyaulacacean dinoflagellate cysts from the British Isles. *Nova Hedwigia Beihefte*, **25**, 579–637.
– (1975). A regional sub-division of dinoflagellate cysts around the British Isles. *New Phytologist*, **75**, 589–603.
– (1978). Dinoflagellate cysts in the plankton. *New Phytologist*, **80**, 219–229.
Reid, P. C., & Harland, R. (1977). Studies of Quaternary dinoflagellate cysts from the North Atlantic. *Contribution series of the American Association of Stratigraphic Palynologists*, **5A**, 147–169.
Reid, P. C., & John, A. W. G. (1978). Tintinnid cysts. *Journal of the Marine Biological Association of the United Kingdom*, **58**, 551–557.
Reinsch, P. F. (1905). Die Palinosphärien, ein mikroskopischer vegetabile Organismus in der Mukronatenkreide. *Centralblatt für Mineralogie, Geologie und Palaeontologie*, 402–407.
Rossignol, M. (1961). Analysis pollinique de sédiments marins Quaternaries en Israel. I. Sédiments récent. *Pollen et Spores*, **3**, 301–324.
– (1962). Analyse pollinique de sédiments marins Quaternaires en Israel. II. Sédiments Pleistocenes. *Pollen et Spores*, **4**, 121–148.
– (1964). Hystrichospheres du Quaternaire en Méditerrané orientale, dans les sédiments Pleistocenes et les boues marines actuelles. *Revue de Micropaléontologie*, 7, 83–99.
Rossignol-Strick, M., & Duzer, D. (1979). Late Quaternary pollen and dinoflagellate cysts in marine cores off West Africa. *"Meteor" Forschungsergebnisse, Reihe C: Geologie und Geophysik*, **30**, 1–14.
Sarjeant, W. A. S. (1961). The Hystrichospheres; a review and discussion. *Grana Palynologica*, **2**, 102–111.
– (1965). The Xanthidia. *Endeavour*, **24**, 33–39.
– (1970). Xanthidia, Palinospheres and 'Hystrix.' A review of the study of fossil unicellular microplankton with organic cell walls. *Microscopy*, **31**, 221–253.
– (1974). *Fossil and Living Dinoflagellates*, pp. 1–182. New York: Academic Press.

– (1978). *Arpylorus antiguus* Calandra, Emend., a dinoflagellate cyst from the Upper Silurian. *Palynology*, 2, 167–178.

Sarjeant, W. A. S., & Downie, C. (1966). The classification of dinoflagellate cysts above generic level. *Grana Palynologica*, 6, 503–527.

– (1974). The classification of dinoflagellate cysts above generic level – a discussion and revisions. *Birbal Sahni Institute of Palaeobotany, Special Publication*, 3, 9–32.

Steidinger, K. A., & Williams, J. (1970). Dinoflagellates. *Memoirs of the Hourglass Cruises*, 2, 1–251.

Stover, L. E., & Evitt, W. R. (1978). Analysis of pre-Pleistocene organic walled dinoflagellates. *Stanford University Publications, Geological Sciences*, 15, 1–300.

Taylor, F. J. R. (1975). Taxonomic difficulties in red-tide and paralytic shellfish poison studies: the "tamarensis complex" of *Gonyaulax*. *Environmental Letters*, 9, 103–119.

– (1976). Dinoflagellates from the International Indian Ocean Expedition. A report on material collected by the R. V. "Anton Bruun" 1963–1964. *Bibliotheca Botanica*, 132, 1–234.

– (1979). The toxigenic gonyaulacoid dinoflagellates. In *Toxic Dinoflagellate Blooms, Proceedings of the Second International Conference on Toxic Dinoflagellate Blooms*, ed. D. L. Taylor & H. H. Seliger, pp. 45–57. New York: Elsevier North-Holland.

– (1980). On dinoflagellate evolution. *BioSystems*, 13, 65–108.

Turpin, D., Dobell, P., & Taylor, F. J. R. (1978). Sexuality and cyst formation in Pacific strains of the toxic dinoflagellate *Gonyaulax tamarensis*. *Journal of Phycology*, 14, 235–238.

Tuttle, R. C., & Loeblich, A. R. (1974). Genetic recombination in the dinoflagellate *Crypthecodinium cohnii, Science*, 185, 1061–1062.

von Stosch, H. A. (1964). Zur Problem der sexuellen Fortpflanzung in der Peridineengattung *Ceratium*. *Helgoländer wissenschaftliche Meeresuntersuchungen*, 10, 140–152.

– (1965). Sexualität bei *Ceratium cornutum* (Dinophyta) *Naturwissenschaften*, 52, 112–113.

– (1972). La signification cytologique de la "cyclose nucleaire" dans le cycle de vie des dinoflagellés. *Memoires Publiciés par la Sociétébotanique de France*, 201–212.

– (1973). Observations on vegetative reproduction and sexual life cycles of two freshwater dinoflagellates. *Gymnodinium pseudopalustre* Schiller and *Woloszynskia apiculata* sp. nov. *British Phycological Journal*, 8, 104–134.

Vozzhennikova, T. F. (1965). *Introduction to the study of fossil peridinian algae*, trans. K. Syers, ed. W. A. S. Sarjeant, pp. 1–231, Boston Spa, England: National Lending Library for Science and Technology.

Walker, L. M., & Steidinger, K. A. (1979). Sexual reproduction in the toxic dinoflagellate *Gonyaulax monilata*. *Journal of Phycology*, 15, 213–315.

Wall, D. (1965). Modern hystrichospheres and dinoflagellate cysts from the Woods Hole Region. *Grana Palynologica*, 6, 279–314.

– (1970). Quaternary dinoflagellate micropaleontology: 1959 to 1969. *Proceedings of the North American Paleontological Convention*, Part G, 844–866.

– (1971). Biological problems concerning fossilizable dinoflagellates. *Geoscience and Man*, 3, 1–15.

– (1975). Modern dinoflagellates as a standard for paleontological inquiry (summary). *American Association of Stratigraphic Palynologists, Contributions Series*, No. 4, 37–43.

Wall, D., & Dale, B. (1966). "Living fossils" in Western Atlantic plankton. *Nature (London)* 211, 1025–1026.
– (1968a). Modern dinoflagellate cysts and evolution of the Peridiniales. *Micropaleontology*, 14, 265–304.
– (1968b). Quaternary calcareous dinoflagellates (Calciodinellidae) and their natural affinities. *Journal of Paleontology*, 42, 1395–1408.
– (1969). The "Hystrichosphaerid" resting spore of *Pyrodinium bahamense*, Plate, 1906. *Journal of Phycology*, 5, 140–149.
– (1970). Living hystrichosphaerid dinoflagellate spores from Bermuda and Puerto Rico. *Micropaleontology*, 16, 47–58.
– (1971). A reconsideration of living and fossil *Pyrophacus* Stein, 1883 (Dinophyceae). *Journal of Phycology*, 7, 221–235.
Wall, D., Dale, B., & Harada, K. (1973). Descriptions of new fossil dinoflagellates from the Late Quaternary of the Black Sea. *Micropaleontology*, 19, 18–31.
Wall, D., Dale, B., Lohmann, G. P., & Smith, W. K. (1977). The environmental and climatic distribution of dinoflagellate cysts in modern marine sediments from regions in the North and South Atlantic Oceans and adjacent seas. *Marine Micropaleontology*, 2, 121–200.
Wall, D., & Evitt, W. R. (1975). A comparison of the modern genus *Ceratium* Schrank 1873 with certain marine dinoflagellates. *Micropaleontology*, 21, 14–44.
Wall, D., Guillard, R. R. L., & Dale, B. (1967). Marine dinoflagellate cultures from resting spores. *Phycologia*, 6, 83–86.
Wall, D., Guillard, R. R. L., Dale, B., Swift, E., & Watabe, N. (1970). Calcitic resting cysts in *Peridinium trochoideum* (Stein) Lemmerman, an autotrophic marine dinoflagellate. *Phycologia*, 9, 151–156.
Williams, D. B. (1971). The occurrence of dinoflagellates in marine sediments. In *Micropalaeontology of Oceans*, ed. B. M. Funnell & W. R. Riedel, pp. 231–243. Cambridge: Cambridge University Press.
Williams, G. L. (1977). Dinocysts. Their classification, biostratigraphy and palaeoecology. In *Oceanic Micropalaeontology*, ed. A. T. S. Ramsay, Vol. 2, pp. 1231–1325. New York: Academic Press.
– (1978). Dinoflagellates, Acritarchs and Tasmanitids. In *Introduction to Marine Micropaleontology*, ed. B. U. Haq & A. Boersma, pp. 293–326. New York: Elsevier North-Holland.
Williams, G. L., Sarjeant, W. A. S., & Kidson, E. J. (1978). A glossary of the terminology applied to dinoflagellate amphiesmae and cysts and acritarchs: 1978 Edition. *American Association of Stratigraphic Palynologists, Contribution Series*, No. 2A, 1–121.
Yentsch, C. M., Lewis, C. M., & Yentsch, C. S. (1980). Biological resting in the dinoflagellate *Gonyaulax excavata. BioScience*, 30, 251–254.
Yentsch, C. M., & Mague, F. C. (1980). Evidence of an apparent annual rhythm in the toxic red tide dinoflagellate *Gonyaulax excavata. International Journal of Chronobiology*, 7, 77–84.
Zederbauer, E. (1904). Geschlechtliche und vergeschlechtliche Fortpflanzung von *Ceratium hirundinella. Deutsche Botanische Gesellschaft, Berichte*, 22, 2–8.
Zingmark, R. G. (1970). Sexual reproduction in the dinoflagellate *Noctiluca miliaris* Suriray. *Journal of Phycology*, 6, 122–126.

Index